Contents

PLANNING PRIMARY
DESIGN &
TECHNOLOGY

LIBRARY

JOHN MURRAY

Other titles in the **Key Strategies** series:

Planning Primary Science Revised National Curriculum Edition by Roy Richardson, Phillip Coote and Alan Wood
Primary Science: A Complete Reference Guide by Michael Evans
Physical Education: A Practical Guide by Elizabeth Robertson
From Talking to Handwriting by Daphne M. Tasker
Helping with Handwriting by Rosemary Sassoon
Planning Primary History by Tim Lomas, Christine Burke, Dave Cordingley, Karen McKenzie and Lesley Tyreman
English Speaking and Listening by Gordon Lamont
Music: A Practical Guide for Teachers by Alan Biddle and Lynn Dolby
Planning Primary Geography by Maureen Weldon and Roy Richardson

Cover photograph: ZEFA

© Roy Richardson 1996

First published in 1996
by John Murray (Publishers) Ltd
50 Albemarle Street
London W1X 4BD

Layouts by Christie Archer
Line drawings by Mark Dimond/Technology Teaching Systems Ltd and Chartwell Illustrators

Typeset in 10½/12 Rockwell by Anneset, Weston-super-Mare, Avon
Printed in Great Britain by St Edmundsbury Press, Bury St Edmunds

A CIP catalogue record for this book is available from the British Library.

ISBN 0-7195-7171-5

Introduction

PLANNING FOR DESIGN AND TECHNOLOGY THROUGH KEY STAGES 1 AND 2

Design and technology is one of the most exciting areas of the primary curriculum, with its emphasis upon providing opportunities for children to design and make objects using a wide range of materials. However, over the past few years there was confusion regarding exactly what was meant by design and technology. The present Orders, statutory as from September 1995, are an obvious attempt to clarify this confusion. The Orders give much clearer guidance, which makes it easier for primary teachers to plan, as well as providing a clear explanation of exactly what makes an activity a design and technology activity.

Quality in any sphere takes time to achieve and to obtain quality you require a concise image of your goals. The teaching of design and technology is no exception.

Where the teaching of design and technology is well established in a school there is a practical link between curriculum areas, particularly within science, mathematics and art. Well planned activities of both short- and long-term duration provide young children with opportunities to link knowledge and skills from other curriculum areas within practical work in 'real life' contexts. Links are often made with local industry or to specific work skills of parents of children within the school, so that children can experience, first hand, the work of designers. In this way, children develop a greater understanding of the ways in which the objects that surround them have been designed and made.

This support material is aimed to help primary schools review, or make a fresh start in implementing design and technology throughout Key Stages 1 and 2. From the start, primary teachers will need a clear picture of exactly what makes a design and technology activity, and the knowledge and confidence to ensure quality of children's learning. The material in this book will assist teachers to plan and deliver this entitlement.

The information provided takes a firm whole-school stance on curriculum development. General guidance is set down from the start for schools and curriculum coordinators. This will need to be considered and addressed prior to beginning the process of curriculum development. Sound initial planning, coupled with a whole-school commitment and understanding of the task are crucial to the success of any curriculum development programme.

Primary schools establishing design and technology within the whole curriculum will need to establish a Policy for Design and Technology which ensures that the statutory requirements are being met, as well as supporting teachers in their planning. A Policy for Design and Technology will allow headteachers and senior management to see more clearly how and when different aspects of the subject will be taught and so provide a way of managing progression throughout the Key Stages.

To assist primary schools in addressing this task, Section 2 shows clearly how schools can set about drawing up a Policy for Design and Technology, irrespective of the size of the school or how children's age groups are arranged within it.

Time is a commodity which primary teachers are finding increasingly in short supply. In order to save teachers' time, a sample Design and Technology Policy (pages 26–30) has been included. Schools may choose to adopt, or adapt, it to suit their school's own situation and ethos.

The sample Design and Technology Policy has been used to create a Scheme of Work for Design and Technology (section 3) that provides a wide range of detailed activities which schools can use or adapt to suit their needs and interests. The Scheme of Work gives design and technology activities for Key Stages 1 *and* 2, which together ensure the full coverage of the statutory Orders. The activities are set down in units of work. Each unit of work has been written to the same format and may be used by teachers wishing to write their own design and technology activities. Also matched to each unit of work are assessment paragraphs, which give examples of the types of response that teachers might observe when children are undertaking a particular unit. Paragraphs have been provided at three levels to give a clearer picture of the range of responses that children might make when undertaking the same unit of work. A unit of work has been provided for each term across both Key Stages 1 and 2, with a balance between units of short and long duration.

Any curriculum development takes time and commitment and schools will need to set down a realistic timetable. Headteachers should ensure that planning for design and technology is given a priority during this period.

This book will give primary schools clear and comprehensive advice in reviewing or drawing up their plans for design and technology in a way that, I hope, will establish the subject as both exciting and motivating for children and teachers alike.

If you have any comments to make on any aspect of the content of this book then I would be pleased to hear from you.

Roy Richardson

Before making a start ...

The role of the headteacher is crucial in any curriculum development and, without their total commitment, support and guidance, it is difficult to achieve a great deal. Headteachers need to ensure that priorities are set and understood by all and are matched against a realistic timetable.

Drawing up a Design and Technology Policy, or deciding what is to be taught, is relatively easy to achieve. The real challenge lies in ensuring that all teachers understand what is required and feel confident and motivated to teach in a way that provides quality learning experiences for the children. In other words, the main aim of curriculum development is not the documentation but the outcome in terms of quality of teaching and learning.

1 PLANNING FOR DESIGN AND TECHNOLOGY

Making a new start

Design and technology is an exciting area of the curriculum for both children and teachers. Unfortunately, many people find this opening statement difficult to support. For the majority, the Orders for design and technology have been unfathomable and primary teachers have devoted a considerable amount of time to interpreting what they were expected to plan, teach and assess. The simplification of the Orders for design and technology now make the requirements clearer and more easily accessed.

The new revised Orders set down what schools should plan and deliver in design and technology. With primary schools already implementing plans for design and technology, it is important that their revised plans set down exactly what is expected. Teachers should feel that they have the skills and knowledge to enable them to implement the Orders, and they should feel excited and motivated at the prospect of working on activities with children.

Coordinators and headteachers with responsibility for leading staff through the planning of the new Orders need to think very carefully about each stage of planning and set a timescale for implementation of the new Orders. Many teachers will have already taught to one set of Orders for design and technology. It is imperative that future plans can be implemented by all teachers and that they feel motivated about teaching the new Orders.

THE STAGES OF IMPLEMENTATION

Headteachers and curriculum coordinators with responsibility for planning to fulfil the new Orders may consider the following stages of implementation;
- **Stage 1 Knowledge of the Orders**
- **Stage 2 Identifying Strengths – carrying out an audit**
- **Stage 3 Writing Units of Work**
- **Stage 4 Mapping Coverage**

Stage 1 Knowledge of the Orders

Before setting out to plan units of work or to adapt existing plans, it is important to ensure that all staff have a clear understanding of the Orders. Explanations should be brief and simple, outlining the main points and, wherever possible, relating them to existing work. This section sets out the Orders in chart form for easy reference for coordinators and those with responsibility for the development of design and technology when leading staff meetings. The charts may be photocopied and given as handouts to staff or used with an overhead projector. Words highlighted in bold are terms which should be emphasised when making presentations to teachers.

The charts do not use the exact wording of the statutory Orders, but provide a summary for reference. However, no changes in meaning have been made.

3

Overview of the new Orders

- A single Programme of Study which sets out what children should be taught within a Key Stage.

- The main aim of design and technology is set out at the beginning of each Programme of Study:

  Pupils should be taught to develop their design and technology capability through combining their Designing and Making skills with Knowledge and understanding in order to design and make products.

- The Programme of Study is set out in the same format within each Key Stage.

- The Programme of Study is set out under five paragraphs:

  PARAGRAPH 1: outlines the types of activities that children should be undertaking.

  PARAGRAPH 2: sets out the **Materials** that children should be given opportunities to use within activities. It also emphasises the need for children
    - to develop understanding of the **working characteristics of materials;**
    - to **work independently** and **in teams** (Key Stage 2);
    - to **apply skills and knowledge from other curriculum areas** within design and technology activities.

  PARAGRAPH 3 sets out the **Designing skills** to be taught.

  PARAGRAPH 4 sets out the **Making skills** to be taught.

  PARAGRAPH 5 sets out the **Knowledge and understanding** to be taught.

- There are two Attainment Targets, called **Designing** and **Making**, which set out the expected standards for pupils' performance.

- Teachers make assessments against **Level Descriptions** that describe the range of performance that pupils working at a particular level should demonstrate.

- When deciding upon a child's particular level, teachers should decide **the best fit level** after considering statements within both Attainment Targets for that level.

PLANNING PRIMARY DESIGN AND TECHNOLOGY

© John Murray

The statutory requirements for Key Stage 1

PARAGRAPH 1

Pupils should be given opportunities to undertake:

- assignments in which they **design and make products**;

- **focused practical tasks** in which they develop and practise skills and knowledge;

- activities in which they **investigate, disassemble and evaluate products**.

PARAGRAPH 2

Pupils should be given opportunities to:

- work with the following materials:
     sheet materials
     reclaimed materials
     textiles
     food
     construction kits;

- **investigate** the working characteristics of materials;

- **apply skills and knowledge from other curriculum areas.**

PARAGRAPH 3

DESIGNING SKILLS

Pupils should be taught to:

- **draw upon experiences** to generate ideas;

- clarify ideas through **discussion**;

- develop ideas through **shaping, assembling and rearranging materials**;

- develop and communicate ideas through **drawing and modelling**;

- **suggest ways of proceeding**;

- **consider their designs** as they develop and **identify strengths and weaknesses**.

The statutory requirements for Key Stage 1 (continued)

PARAGRAPH 4

MAKING SKILLS

Pupils should be taught to:

- **select materials, tools and techniques;**
- **measure, mark out, cut and shape** a range of materials;
- **assemble, join and combine** materials and components;
- apply **finishing techniques;**
- **suggest how to proceed;**
- **evaluate** their products as they develop, **identifying strengths and weaknesses.**

PARAGRAPH 5

KNOWLEDGE AND UNDERSTANDING

Pupils should be taught:

mechanisms
- to use simple mechanisms, including wheels and axles;

structures
- to make structures more stable and withstand greater loads;

products and applications
- to investigate and disassemble simple products in order to learn how they function;
- to relate the ways things work to their intended purpose, how materials and components have been used, people's needs, and what users say about them;

quality
- that the quality of a product depends on how well it is made and how well it meets its purpose;

health and safety
- knowledge of health and safety, as consumers and when working with materials and components, including:
  - considering the hazards and risks in their activities;
  - following simple instructions to control risk to themselves;

vocabulary
- to use appropriate vocabulary for naming and describing equipment, materials and components they use.

The statutory requirements for Key Stage 2

PARAGRAPH 1

Pupils should be given opportunities to undertake:

- assignments in which they **design and make products**;

- **focused practical tasks** in which they develop and practise skills and knowledge;

- activities in which they **investigate, disassemble and evaluate products.**

PARAGRAPH 2

Pupils should be given opportunities to:

- work with the following materials:
      stiff and flexible materials
      framework materials
      mouldable materials
      textiles
      food
      electrical components
      mechanical components
      construction kits;

- **work independently and in teams;**

- **apply skills and knowledge from other curriculum areas.**

DESIGNING SKILLS

PARAGRAPH 3

Pupils should be taught to:

- **use information sources to help their designing;**

- **generate ideas, considering the users and purposes** for their designs;

- **clarify ideas, develop criteria** for their designs and **suggest ways forward**;

- **consider appearance, function, safety and reliability** when developing proposals;

- explore, develop and communicate aspects of their design proposals by **modelling their ideas in a variety of ways;**

The statutory requirements for Key Stage 2 (continued)

- develop a clear idea of what has to be done, **proposing a sequence of actions and suggesting alternative methods** of proceeding if things go wrong;

- **evaluate their design ideas as they develop**, bearing in mind the users and the purposes for which the product is intended, and **indicate ways of improving their ideas**.

MAKING SKILLS PARAGRAPH 4

Pupils should be taught to:

- **select appropriate materials, tools and techniques;**

- **measure, mark out, cut and shape** a range of materials, using additional tools, equipment and techniques;

- **join and combine materials** and components accurately **in temporary and permanent ways**;

- **apply additional finishing techniques** appropriate to the materials being used and the purpose of the product;

- **develop a clear idea of what has to be done, planning how to use materials, equipment and processes, and suggesting alternative methods of making if first attempts fail;**

- **evaluate** their products, **identifying strengths and weaknesses**, and **carrying out appropriate tests**;

- **implement improvements that they have identified.**

KNOWLEDGE AND UNDERSTANDING PARAGRAPH 5

Pupils should be taught:
materials and components

- how the characteristics of materials relate to the ways they are used;

- how materials can be combined and mixed to create more useful properties;

control
- how simple mechanisms can be used to produce different types of movement;

- how electrical circuits, some with simple switches, can be used to achieve functional results;

The Statutory requirements for Key Stage 2 (continued)

structures

- how structures can fail when loaded, and techniques for reinforcing and strengthening them;

products and applications

- to investigate, disassemble and evaluate simple products and applications, including those with mechanical and electrical components, in order to learn how they function;

- to relate the ways things work to their intended purpose, how materials and components have been used, people's needs, and what users say about them;

quality

- to distinguish between how well a product has been made and how well it has been designed;

- to consider the effectiveness of a product, including the extent to which it meets a clear need, is fit for its purpose and uses resources appropriately;

health and safety

- further knowledge and understanding of health and safety, as designers, makers and consumers, including:
  - recognising hazards to themselves and to others in a range of products, activities and environments
  - assessing risks to themselves and to others
  - taking action to control these risks;

vocabulary

- to use appropriate vocabulary for naming and describing equipment, materials and components, and processes they use.

Attainment Targets – Level Descriptions

In deciding upon a child's level of attainment teachers should decide which description best fits.

DESIGNING ATTAINMENT TARGET 1

Level 1   When designing and making, pupils generate ideas through shaping, assembling and rearranging materials and components. They recognise the simple features of familiar products and, when prompted, relate them to their own ideas. They use pictures and words to convey what they want to do.

Level 2   When designing and making, pupils use their experience of using materials, techniques and products to help generate ideas. They use models and pictures to develop and communicate their designs. They reflect on their ideas and suggest improvements.

Level 3   When designing and making, pupils generate ideas, recognising that their designs will have to satisfy conflicting requirements. They make realistic suggestions about how they can achieve their intentions and suggest more ideas when asked. They draw on their knowledge and understanding of the appropriate Programme of Study to help them generate ideas. Labelled sketches are used to show the details of their designs.

Level 4   When designing and making, pupils gather information independently, and use it to help generate a number of ideas. They recognise that users have views and preferences, and are beginning to take them into account. They evaluate their work as it develops, bearing in mind the purposes for which it is intended. They illustrate alternatives using sketches and models and make choices between them, showing an awareness of constraints.

Level 5   When designing and making, pupils generate ideas that draw upon external sources and their understanding of the characteristics of familiar products. They clarify their ideas through discussion, drawing and modelling, using their knowledge and understanding of the appropriate Programme of Study to help them. Pupils evaluate ideas, showing understanding of the situations in which their designs will have to function, and awareness of resources as a constraint.

Attainment Targets Level – Descriptions

MAKING **ATTAINMENT TARGET 2**

Level 1 When designing and making, pupils explain what they are making and which materials they are using. They select from a narrow range of materials and use given techniques and tools to shape, assemble and join them.

Level 2 When designing and making, pupils select from a range of materials, tools and techniques, explaining their choices. They manipulate tools safely and assemble and join materials in a variety of ways. They make judgments about the outcomes of their work.

Level 3 When designing and making, pupils think ahead about the order of their work, choosing tools, materials and techniques more purposefully. They use tools with some accuracy and use simple finishing techniques to improve their products. They cut and shape materials and components with some precision, to help assembly. Their products are similar to their original intentions and, where changes have been made, they are identified.

Level 4 When designing and making, pupils produce step-by-step plans that identify the main stages in making, and list the tools, materials and processes needed. They measure, mark out and cut simple forms in a variety of materials and join them using a range of techniques. They show increasing accuracy, paying attention to quality of finish and function. They identify what is, and what is not, working well in their products.

Level 5 When designing and making, pupils work from plans they have produced, modifying them in the light of difficulties. They use a range of tools, materials and processes safely and with increasing precision and control. They use measuring and checking procedures as their work develops, and modify their approach if first attempts fail. They evaluate their products by comparing them with their design intentions and suggest ways of improving them.

Stage 2 Identifying strengths

CARRYING OUT AN AUDIT

Schools should aim to write a Scheme of Work for design and technology that builds on the good practice that exists within the school, the quality design and technology activities that are already undertaken and the skills, knowledge and interests of the staff.

The following points should be considered before commencing the time consuming and lengthy process of drawing up plans for the implementation of the new Orders:

- **What design and technology is being taught in the school at present?**
  It is important to first look at present plans and identify work that is already being undertaken and that the staff feel confident with and enjoy teaching. Some changes may be necessary to ensure that the work applies to the new Orders but good design and technology activities will require minimal changes. Many schools, particularly at Key Stage 1, arrange a visit to a place of interest in the locality and then create an interesting environment within the classroom, with the children making items for inclusion. For instance, schools often change the environment of a corner of the classroom to a post office, hairdressers or whatever fits in with other work being undertaken. Such work provides many opportunities for children to design and make products.

- **What are the school's strengths?**
  Identify the school's strengths in design and technology. If, for example, the school is particularly good at textile work, ensure that enough units of work are planned to build upon and exploit the established skills in a design and technology context.

- **What are the school's weaknesses?**
  If the school has weaknesses in any particular area, time should be provided for developing specific skills.

- **What resources are available?**
  Take an audit of the resources and equipment available and decide the areas in which the school is adequately equipped for design and technology. Make a note of areas in which the school is under resourced and make a prioritised list of items the school will need to acquire in the future. Resource requirements should be set down within the school's short- and long-term plans.

- **Are you equipped to undertake work with food?**
  The Orders do not require schools to equip themselves with cookers and purpose-built food technology areas. If the school already has such provision, this should be viewed as an asset and incorporated in a number of planned units of work. If the school does not have such equipment, units of work can be planned that have food as the focus but do not require new equipment. Design and technology coordinators should not make excessive financial demands upon schools in order to undertake work on food.

The units of work dealing with food within the Scheme of Work have been specifically planned for schools with little or no specific food equipment.

■ **Are you addressing the health and safety aspects of working with food?**

Schools undertaking any work where children will be tasting food will need to address the following:

Children working with food

Establish a set of rules for all children to follow whenever they work with food. Ensure that all staff are fully aware of the rules, which should be on display for children to consult.

Special diets

A letter should be sent to parents informing them that their child will be tasting food. They will need to inform you if their child follows any special diet that could prevent them from tasting certain foods. Records should be kept and all teachers should have easy access to such information.

Equipment

Equipment for food-technology work will need to be purchased, stored and marked for food-use only.

Suitable cleaning materials, including a steriliser, should be available. This is particularly true where no specific food area exists and work has to be undertaken in the classroom.

In many schools the siting of a cooker, even a mobile one, can prove difficult. However, a cooker is not essential as there are many activities that can be undertaken without children having to heat food. If you do not have a cooker then think carefully before purchasing one and consider the degree to which the cooker will be used as opposed to the financial outlay.

Clothing

Aprons, specifically for children to use when working with food, should be available.

Setting out a food area

Some schools may have a food area while others do the work in classrooms. In either case procedures should be established for the setting out of the working area. In the classroom plastic sheets, specifically for using within food activities, should be available together with a suitable steriliser for cleaning down before and after use.

All staff should be fully aware of the procedures for setting out and clearing away food areas. The procedures should also be explained to the children and they should be encouraged to take on a greater degree of responsibility for their work areas as they progress through the school.

Decide who will take responsibility for cleaning tea towels and dishcloths so that they are available for use in the next lesson.

Storage of food

Your Scheme of Work will indicate the types of food activities you are going to undertake and provide guidance on the types of foods to be used. If perishable food is to be stored a fridge is essential.

Further guidance

Consult your Local Authority for any guidance they might provide.

Read the relevant sections in the school's Health and Safety Manual.

For details on all aspects of food technology in primary schools consult: *Working with Food in the Primary School* by Jenny Ridgwell, published by Ridgwell Press.

■ **What are the staff's skills and interests?**
It is important to consider the areas in which the staff are particularly skilled or interested, as they are much more likely to teach a unit well if it is about a subject with which they are familiar. It is important that teachers develop a sound knowledge of any area that they are asked to teach.

■ **How can use be made of the school's locality?**
It is helpful to know the locality and to consider what it has to offer in terms of supporting the teaching of design and technology. Examples of how the local environment might be put to good use are:
– A visit to a local shop to observe how it functions, with a view to creating a similar environment in the classroom.
– A visit to a local craftsperson who can demonstrate working with a particular material. For example, a potter may well be able to demonstrate techniques that children could use to create a soap dish.
– A visit to the nearby industrial estate to observe a printer at work. For example, observing how a printer creates labels for products could lead to children applying the ideas observed to making their own labels for healthy food bars they have made.

■ **What role can parents and adult helpers play?**
The most obvious role for parents and adult helpers is in supporting teachers in the classroom. The practical nature of design and technology means that it makes considerable demands upon the teacher when a large number of children are designing and making. Parents can be shown how to support specific activities such as working with clay or textiles.

Parents may have specific skills which can be demonstrated in the classroom. A parent may be skilled at working with certain materials and could assist children in developing this skill in a design and technology context.

Parents can provide valuable links with local industries. They may be keen to allow children to visit their workplace, to talk to the children or to sponsor a design and technology project linked to their work by providing equipment, materials or finances. For example, a firm that chose to sponsor the school's football and netball kits would be happy to see that the school had used this opportunity for one class to design and make a package to hold the kits so that they could be displayed at the school entrance.

■ **Is there room for impromptu design and technology activities?**
A good primary school is well planned and organised and is a place where a variety of activities and events occur each term. Primary schools have always been good at exploiting the impromptu moment when the unexpected happens. Planning should be tight, providing all children with opportunities to design and make with a variety of materials, but it should also allow teachers the opportunity to work spontaneously.

■ **Have the teachers sufficient time to teach design and technology?**
Look carefully at each unit of work that is planned and consider the amount of time it will take to complete. Can the unit be completed by all children in the time available? It is important to ensure that teachers do not feel under pressure to complete a task.

■ **How can quality be achieved?**
All teachers should be aware that one of their main tasks is to ensure that children are involved in making articles of quality. Children need to consider quality from the outset through each stage of making. However, it is the teachers who set the standards with regard to the quality of anything that children design and make.

■ **How will work be assessed?**
It is important to give thought to the demands of assessment. Addressing what might be assessed at the end of each unit clarifies how the unit might be planned and differentiated, as well as highlighting what aspects would contribute to assessments.

All teachers should be encouraged to identify what they might assess against each unit of work, to aid their planning and teaching.

■ **How will differentiation and progression be addressed?**
Each planned unit of work should address differentiation to ensure that activities are matched to children's abilities and build upon previous experience. The units of work set out in Section 3 provide guidance but can be adapted to suit teachers' needs.

Coordinators should read through the units of work they have drawn up to ensure that differentiation has been addressed at the planning stage and that they provide a progressive route through Key Stages 1 and 2.

Stage 3 Writing units of work

Schools will need to draw up a unit of work for each term throughout Key Stages 1 and 2 (eighteen units of work in total).

To assist schools, a blank unit of work planning sheet has been provided in Appendix 1. Below is a copy of a blank planning sheet and a sample completed planning sheet.

The sample gives examples of the type of responses that teachers may observe as the children work through this particular unit. They provide support and guidance for teachers when planning to teach this particular unit of work or when adapting it to match their own school situation. They also provide guidance for teachers when undertaking assessment in design and technology.

Teachers may well observe very different sets of responses but should find that the information provided helps them to gain an understanding of the types of responses that contribute towards forming a clearer picture of a child's design and technology capability.

KEY STAGE 1
UNIT OF WORK 2

SOUVENIRS

Context As part of a school visit children can take the opportunity to look round a souvenir shop. The teacher can bring to their attention the range of souvenirs but particularly ones that present the children with ideas for designing and making with the range of materials they have available. They could look at bookmarks or clay pots that could be used as the basis for developing their own design ideas. The teacher should return to school with a number of souvenirs that can be used to stimulate children's thinking.

Outcome Making a souvenir for an area or place that they have visited.

Cross-curricular links The work helps to develop children's understanding of a contrasting locality.
GEOGRAPHY

RESOURCES FOR INVESTIGATING
■ A collection of souvenirs purchased from a place visited on a school trip. The collection should include a number of souvenirs that the children can make for themselves – e.g. a bookmark.

RESOURCES FOR MAKING
■ This will depend upon the types of souvenirs that you have collected and the materials that you have decided will form the focus for this unit of work. Choose a collection of souvenirs that will focus the children's attention upon certain materials – e.g. textiles. This is a particularly good unit for working with textiles. Children may decide to make bags, bookmarks, belts or badges.

ACTIVITIES   **Focused practical task**

■ Provide opportunities for the children to learn how to work with textiles. They should be able to join through sewing and sticking and produce simple sewing patterns.
■ Teach the children how to decorate their textiles by sticking and sewing other materials such as buttons, sequins etc.
■ Demonstrate how to sew two pieces of material together and to stuff the insides to produce a soft toy effect.
■ Demonstrate and teach the skills and tools involved in making the souvenirs that you have collected.

Investigative, disassembly and evaluative activities

■ Make a display of the souvenirs and other items brought back from the school trip. It is a good idea to include photographs in the display, as these will provide children with ideas.
■ Draw the children's attention to the different souvenirs and talk about the materials from which they are made.
■ Ask the children how they know that these are souvenirs.
■ On a sheet of drawing paper, ask the children to draw one of the souvenirs and around it write or draw everything they know about it. They can, for

36

Figure 1.1 A completed unit of work sheet

Appendix 1 – Unit planning sheet

Year/s	Term	Time allocation
Unit of work		
Title		

Context

Outcome

Cross-curricular links

RESOURCES FOR INVESTIGATING

RESOURCES FOR MAKING

ACTIVITIES   **Focused practical task**

Investigative, disassembly and evaluative activities

118

PLANNING PRIMARY DESIGN AND TECHNOLOGY   © John Murray

Figure 1.2 A blank unit of work sheet (Appendix 1)

GUIDANCE ON WRITING YOUR OWN UNIT OF WORK

This section below provides guidance for teachers in completing a blank unit of work planning sheet or interpreting or adapting those provided in this book. Guidance is provided under each heading. Once completed, the units of work can be combined to make up the school's Scheme of Work.

Context This indicates whether the project is part of a major topic for the term or forms an integral part of a school visit.

Outcome Gives an indication of what the children are going to make.

Cross-curricular links Shows where the unit of work has a direct link with another area of the curriculum. Design and technology makes many links with science, mathematics and art.

Resources for investigating This indicates the equipment or materials that the teacher should provide in order to give the children opportunities to investigate, disassemble and evaluate, increasing their knowledge and understanding of what they are going to design and make.

Resources for making This lists the main resources that the teacher needs to provide for the children to use when designing and making. Not all the suitable resources are listed and children may want to add equipment and materials which are readily to hand or that they have experience of working with.

ACTIVITIES Focused practical task This lists the skills that the teacher will need to teach or the activities that children need to undertake in order to fulfil their designing and making tasks. Undertaking these activities prior to making the artefact will save teachers having to teach the skills during the making stage. For some units, teachers will find that this section needs to be addressed first; in others they will find that the investigate, disassembly and evaluation section will need to be undertaken first.

Investigative, disassembly and evaluative activities This lists the activities that the children could undertake in investigating, disassembling and evaluating products in order to increase their knowledge and awareness of what they are going to be asked to design and make.

Design and make assignment Listed are details of the design and make activity which children will undertake. The activity is set out in detail and in the sequence that teachers are advised to address each stage. Listed are important points which should be brought to the children's attention, discussion points and an indication of the quality to be achieved. Details are given of how the work might be evaluated and recorded.

Where an activity needs to be adapted for children with special needs, teachers should set down the details in this section when writing or adapting their own units.

Extension activities Listed here are activities relating to the main design and make task that children may move on to undertake.

Also listed are activities which may be undertaken at a later date as a result of the knowledge and skills acquired in this activity. The activities listed in this section provide further ideas for teachers.

Assessment Examples are provided of assessment opportunities that are specific to the unit of work. Examples have been provided for levels 1 to 3 for Key Stage 1 units of work and for levels 3 to 5 for Key Stage 2 units of work. It is however, recognised that children at both Key Stages may be working outside these levels.

Stage 4 Mapping coverage

When all the units of work have been written, they should be checked to ensure that, together, they provide coverage of the statutory Orders at each Key Stage. To assist schools in mapping coverage across the Key Stages, a blank mapping sheet for each Key Stage has been provided in the Appendix.

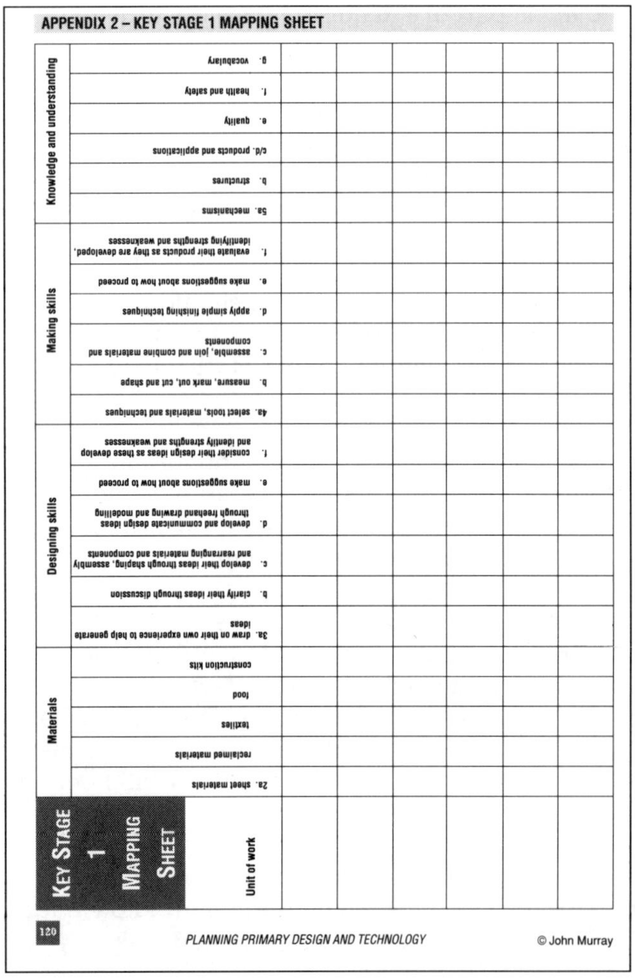

APPENDIX 2 – KEY STAGE 1 MAPPING SHEET

KEY STAGE 1 MAPPING SHEET — Unit of work

Knowledge and understanding
- g. vocabulary
- f. health and safety
- e. quality
- c/d. products and applications
- b. structures
- 5a. mechanisms

Making skills
- f. evaluate their products as they are developed, identifying strengths and weaknesses
- e. make suggestions about how to proceed
- d. apply simple finishing techniques
- c. assemble, join and combine materials and components
- b. measure, mark out, cut and shape
- 4a. select tools, materials and techniques

Designing skills
- f. consider their design ideas as these develop and identify strengths and weaknesses
- e. make suggestions about how to proceed
- d. develop and communicate design ideas through freehand drawing and modelling
- c. develop their ideas through shaping, assembly and rearranging materials and components
- b. clarify their ideas through discussion
- 3a. draw on their own experience to help generate ideas

Materials
- construction kits
- food
- textiles
- reclaimed materials
- 2a. sheet materials

120 PLANNING PRIMARY DESIGN AND TECHNOLOGY © John Murray

Figure 1.3 Blank mapping sheet for Key Stage 1

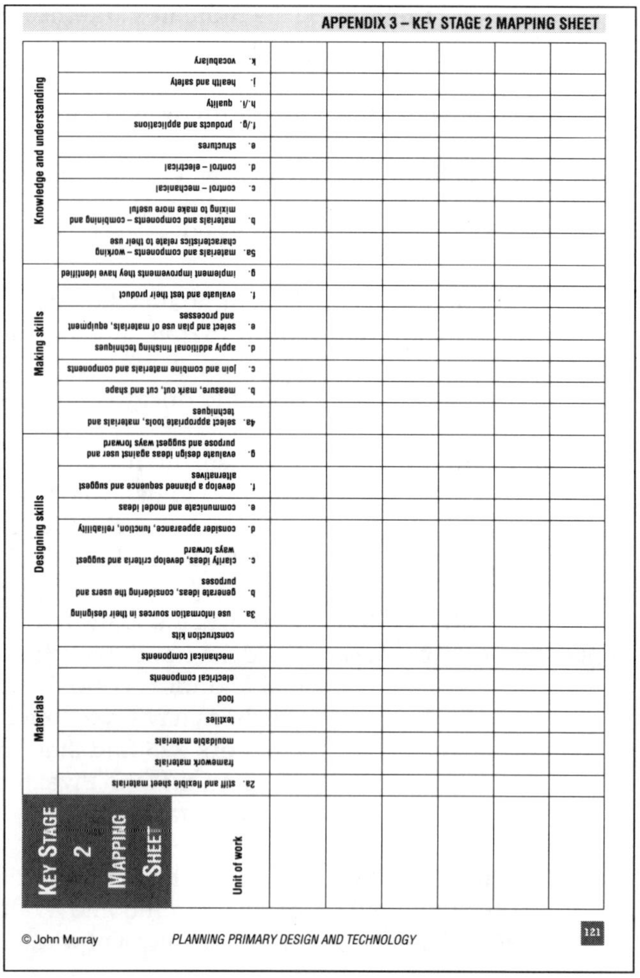

APPENDIX 3 – KEY STAGE 2 MAPPING SHEET

KEY STAGE 2 MAPPING SHEET — Unit of work

Knowledge and understanding
- k. vocabulary
- j. health and safety
- h./i. quality
- f./g. products and applications
- e. structures
- d. control – electrical
- c. control – mechanical
- b. materials and components – combining and mixing to make more useful
- 5a. materials and components – working characteristics relate to their use

Making skills
- g. implement improvements they have identified
- f. evaluate and test their product
- e. select and plan use of materials, equipment and processes
- d. apply additional finishing techniques
- c. join and combine materials and components
- b. measure, mark out, cut and shape
- 4a. select appropriate tools, materials and techniques

Designing skills
- g. evaluate design ideas against user and purpose and suggest ways forward
- f. develop a planned sequence and suggest alternatives
- e. communicate and model ideas
- d. consider appearance, function, reliability
- c. clarify ideas, develop design criteria and suggest ways forward
- b. generate ideas, considering the users and purposes
- 3a. use information sources in their designing

Materials
- construction kits
- mechanical components
- electrical components
- food
- textiles
- mouldable materials
- framework materials
- 2a. stiff and flexible sheet materials

© John Murray PLANNING PRIMARY DESIGN AND TECHNOLOGY 121

Figure 1.4 Blank mapping sheet for Key Stage 2

Each mapping sheet sets out the materials, designing and making skills and knowledge and understanding that must be addressed at each Key Stage. Schools will need to map out the focus for each unit of work and then to see where there may be gaps where aspects of the Orders have not been planned for. In such cases, units of work need to be adapted or new ones written.

All the units of work provided in Section 3 have already been set down on mapping sheets to assist schools in their planning (see pages 32 and 51–52).

2 WRITING YOUR DESIGN AND TECHNOLOGY POLICY

What to consider when writing a policy

The writing and review of a Design and Technology Policy will encourage professional debate among staff and will increase their awareness of what is required from their teaching.

A policy will help to develop continuity and progression and provide guidance for maintaining and developing quality in teaching and in the children's work. It should give guidance to new staff clarifying the school's expectations. A policy should also guide the headteacher, coordinator and any visitors in what to look for when evaluating the quality of the design and technology teaching taking place in the school.

A good curriculum Policy for Design and Technology should:
■ help the class teacher to plan to deliver the National Curriculum;
■ reflect the ideas and philosophy which are promoted in the whole-school curriculum policy for the school;
■ reflect the ethos for the teaching and learning of design and technology capability at the school;
■ guide the class teachers in planning a range of design and technology activities for all children;
■ help to achieve consistency of the design and technology ethos throughout the school;
■ be concise and written in straightforward language that can be understood by parents and governors as well as staff;
■ reflect the work currently being undertaken in the classroom or the work which the whole staff are working towards achieving;
■ direct the teacher to any other policies that need to be considered when planning, for example, Health and Safety Policy or regulations regarding use of glue guns or food preparation and equipment.

Once the staff have an understanding of what constitutes good design and technology practice and are aware of what the statutory Orders require them to plan to cover, it is possible for them to make policy decisions relating to how and when the children will be introduced to the different aspects of National Curriculum design and technology. When policy decisions have been made, it is easier for schools to identify how and when the various parts of the statutory Orders will be taught.

WHAT MAKES A GOOD POLICY?

In order to create a good policy:
■ keep it short and interesting so that people will want to read it and be able to understand it easily;
■ decide on the readership, for example, governors, teachers, parents, visitors to the school and inspectors and advisers;

- ensure it is free from unnecessary design- and technology-specific terminology, especially if it is going to be distributed among non-teaching staff such as parents and governors;
- make the meaning clear, so that a teacher new to the school, or a supply teacher will be able to understand fully what is expected of them;
- clearly show what the school is setting out to achieve so that achievements can be identified and measured.

A good policy is a management tool which can assist headteachers in bringing about change in the quality of education within their school.

A good indicator of how well a policy has been prepared and written is to consider whether a visitor to the school could read the policy and, if everybody were teaching design and technology on that day, could clearly observe aspects of the policy being addressed, worked towards or set down in the school's long-term plan.

DRAWING UP A DESIGN AND TECHNOLOGY POLICY

Set out below are headings which form the framework for a school's Design and Technology Policy. Under each of the headings there is a series of questions designed to help staff to address issues in the drawing up of a policy. Staff should be led through a series of staff meetings in which they are given the opportunity to discuss the issues. Staff should agree the topics to be addressed. They may think also of other issues which are particularly relevant to the school.

At the end of this section is a sample Design and Technology Policy which schools may use as a guide. Coordinators and headteachers have found that the sample policy helps them to plan and guide the discussion at staff meetings. Schools may, of course, choose to adapt and use the sample policy to meet their own situation and requirements. This may be the case in schools where there is insufficient time available to hold a series of discussion meetings with staff.

AREAS TO BE CONSIDERED

Background How has the policy been drawn up? Were all staff involved? Have all staff agreed to its content?

Who was responsible for the final writing up of the policy?

What was the role of the coordinator in drawing up the policy?

How are staff to be made aware of the content of the policy? How does the school ensure that new staff and supply teachers are aware of the policy?

Where can copies of the policy be found? Are all staff provided with copies?

Are parents given access to the policies? Is any other body/person allowed to request a copy of the policy? What are the procedures for requesting a copy of the policy?

Is there any documentation that should be attached as an Appendix to the policy, for example, the school's Policy for Design and Technology, resource lists, school policy on school visits, the Health and Safety Policy or extracts from health and safety guidance on the use of glue guns or undertaking food activities within classrooms?

The philosophy of design and technology

What do the staff feel that design and technology is about? What are the main aspects of design and technology that the children in the school should have experienced and learnt about by the time they leave?

How should design and technology be taught within the school? Should emphasis be given to any particular aspect of design and technology?

Which units of work or aspects of design and technology will be undertaken at Key Stage 1?

Which units of work or aspects of design and technology will be undertaken at Key Stage 2?

Will the policy refer to the Programme for Design and Technology so that details of what is to be covered at each Key Stage can be kept to a minimum in the policy?

Design and technology in the National Curriculum

Do the staff have a broad understanding of the National Curriculum document for design and technology? For example:
– Could staff write a short paragraph which explains the range of materials that children should have the opportunity to have worked with at Key Stages 1 and 2.
– Could they write a paragraph setting down the designing and making skills and the knowledge and understanding children should be taught at each Key Stage?
– Could they explain the statutory requirements for each Key Stage?

Will the policy include details of anything to be emphasised at each Key Stage? It may be that the school decides to emphasise quality at each stage in all design and technology activities.

Has the school drawn up a Policy for Design and Technology which specifies exactly when various aspects of the statutory Orders are to be focused upon within each unit of work? These specifications may be attached to the policy as an Appendix.

Teaching strategies and planning

What guidance do teachers have to assist their planning for design and technology? Is there a Scheme of Work for design and technology? How will the Scheme of Work assist teachers in planning and teaching?

How is design and technology planned within the school?

What approach will staff be required to take in the investigative, disassembly and evaluative aspects of each unit of work?

Will emphasis be given to the development of independence in the way in which children set out their working areas and work at their designing and making?

Are there any specific cross-curricular links? Are these indicated in the Scheme of Work?

Skills and techniques overlap between art and design and technology. How is this to be reflected in the school's Scheme of Work and teachers' planning? Will the skills and techniques required to be taught within a design and technology unit of work be addressed in art lessons?

Are there any aspects that the school wishes to make central to the teaching of design and technology? Is emphasis given to firsthand experience, practical activities, the production of a product and quality designing and making?

Is an investigative approach to design and technology emphasised?

Are there specific links with the local community that need to be emphasised? Is there a particular industry that plays a leading role within the community which should be involved in design and technology studies wherever possible? Have specific links been established and set down within the Scheme of Work?

Does the school take the children on a residential field trip each year? Are there any other trips that classes take each year that have a design and technology focus? Does the school plan to have a design and technology week or day where the whole school works together on design and technology activities?

How are designing and making skills to be developed?

Does the school have any computer software or hardware that is particularly relevant to the teaching of design and technology?

In the classroom

Will there be any specialist teaching for design and technology? Will any specific aspects of design and technology be taught by somebody with a specialist knowledge?

Are there staff with particular strengths in design and technology? Will these assets be utilised by the school?

Is a range of teaching styles utilised, adapted to whatever is being taught? Is this identified within the Scheme of Work?

What types of practice would the school expect to see when design and technology is being taught?

What is meant by investigative design and technology in terms of how design and technology is taught in the classroom?

What role will ancillary staff and other adult helpers have in supporting the teaching of design and technology? Are teachers required to supply guidance to support staff to ensure that aims and objectives for lessons are clearly and fully supported?

Equal opportunities and special needs

How can it be ensured that all children are given the same opportunities within their design and technology work?

How will teachers plan to ensure that all children are given opportunities to work to their full potential, regardless of ability?

How is work to be differentiated? Is this to be made explicit within the Scheme of Work, in teachers' own plans or both?

How are staff expected to address the teaching of children with special needs?

Does design and technology have any specific role in the breakdown of stereotyping within our and other cultures?

Do the units of work for design and technology appeal equally to both boys and girls? Do they support gender stereotyping? Is guidance to be provided on the roles boys and girls take when undertaking work on food or textiles?

Do the school's plans for design and technology develop a greater understanding of other customs and cultures?

Assessment and record keeping

How and when are teachers expected to assess children's progress in design and technology?

What other guidance exists in school on assessment?

How are the Level Descriptions to be used in undertaking assessments?

Does any guidance exist on interpreting the Level Descriptions?

Does the school have a portfolio of children's work that offers guidance on making assessments in design and technology? With most of the children's work being 3D, has the school made provision for work to be photographed to add to the children's assessment portfolios?

As the Level Descriptions are for assessments at the end of each Key Stage, has the school made any provision for making assessments midway through Key Stage 2?

Does the school provide guidance on the types of statements that teachers are expected to make on reports to parents?

Has the school provided any guidance on what the majority of children should be expected to know and be able to do at the end of each Key Stage, in order to aid teachers making assessments?

Resources

How are design and technology resources to be organised within the school?

Who is responsible for the resources, the drawing up of the resource list and the ordering of new equipment? How are teachers to order design and technology equipment?

Are resources to be kept centrally as well as in each classroom? Are the arrangements different for each Key Stage?

Is there a system adopted for the borrowing and return of central resources?

How will children be encouraged to use the resources? Is there guidance on whether children may use certain equipment – e.g. strimmers, glue guns and pointed scissors – and, if so, how they should be used safely?

What aspects of IT will contribute to the teaching of design and technology? When and within which units of work will each aspect of IT be most relevant? Is this information set down in the Scheme of Work?

Where can teachers find a list of IT equipment suitable for supporting the teaching of design and technology?

Early years

How will very young children be introduced to design and technology and the way of working? How will the school ensure that the work they undertake leads in to Key Stage 1 and is built upon thereafter? What are the links between the work of reception aged children and those at Key Stage 1?

How does the teaching in the nursery school contribute to the teaching of design and technology later?

Which materials and tools are young children likely to have had experience of using as a basis for future design and technology activities?

Safety and care What school guidance exists on taking children out of school or on visits? Where can this information be found?

Where is the school's Health and Safety Policy kept?

Are there any school guidelines on any aspect of the teaching of design and technology with particular reference to:
– working with food;
– use of glue guns;
– working with specific items, such as strimmers, photocopiers and pointed scissors;
– use and abuse of glues;
– teaching children correct and safe ways of working with tools;
– giving children direct instructions on any specific health and safety rules.

Will the policy indicate any specific health and safety issues relating to the use of craft knives and glue guns in classrooms?

Working with food What guidance has the school and/or Local Authority provided for teachers on health and safety matters?

How will non-teaching staff be made aware of the health and safety procedures?

Has the school received, from parents, information regarding children's dietary needs? Where is this information available?

Have any rules been established for children to follow when working with food?

Has more specific guidance been set down with regard to children undertaking food activities? Where would this information be found?

When undertaking food activities the appropriate health and safety procedures must be adhered to. Where will teachers find these procedures set down?

When working with food, children should be fully aware of the school's rules regarding personal hygiene. Does the school have a set of rules and where are they set down?

Has the school established any classroom procedures to guide teachers and children and to establish a more consistent approach to food activities as children move through the school?

Review How often will the school review the Design and Technology Policy in order to update and refine it as appropriate?

How will the policy be presented to the governing body? Will it be in the form of a written statement, or will it be supported by workshops, examples of children's work, equipment etc. to add clarity to the school's aims?

A sample Design and Technology Policy

On the following pages is a sample Design and Technology Policy based on the questions outlined on pages 20 to 24.

This sample policy will form the basis upon which the whole of the statutory Orders for design and technology are mapped out in Section 4. It would influence how a school might plan and teach design and technology and would inform new teachers of what the school's expectations are with regard to the teaching of design and technology. The sample policy will assist teachers in drawing up the Scheme of Work and will be clearly reflected throughout it.

The ultimate test of the success of a policy is that it is reflected within the practice that is observed throughout the school.

Background

1. This policy outlines the purpose, nature and management of the design and technology taught and learned in our school. Design and technology is a foundation subject within the National Curriculum.

2. The school Design and Technology Policy reflects the agreed views of all the staff of the school. It has been drawn up as a result of a series of whole-staff meetings led by the design and technology coordinator. The policy has the full agreement of the governing body and staff of the school. The policy was agreed by the governing body at their meeting of (*date*).

3. All staff have read and agreed the Design and Technology Policy and are fully aware of their role in its implementation. All staff have a copy of the policy in their school policy folders.

4. All new members of staff are provided with policy folders and curriculum coordinators have the responsibility for explaining the teaching implications of their own policies.

5. Copies of all policies are kept in folders in the Secretary's office and the headteacher's room. Parents wishing to see a copy of the policy can do so by making a request to the headteacher.

   The following are attached to this policy:
   - **Appendix 1** – the school's outline for teaching Design and Technology which maps out each unit of work. Together, the units of work for each Key Stage ensure total coverage of the National Curriculum;
   - **Appendix 2** – the school's design and technology resource list;
   - **Appendix 3** – a copy of the school's Health and Safety Policy;
   - **Appendix 4** – health and safety guidance specific to aspects of the teaching of design and technology.

The philosophy of design and technology

1. Design and technology is about children developing designing and making skills that they can combine with specific knowledge and understanding in order to design and make quality products. The process should assist children in developing a greater awareness and understanding of how everyday products and items are designed and made. Throughout their compulsory education, children will become increasingly aware of the technological contribution made to both our culture and quality of life.

2. Design and technology draws upon knowledge and skills from other curriculum areas, particularly mathematics, science and art.

3. All children will undertake a design and technology activity each term. All design and technology activities will involve children being taught the relevant skills, techniques and knowledge required to undertake the designing and making aspects with increased confidence and knowledge.

4. Prior to designing and making, all children will be provided with opportunities to investigate, disassemble and/or evaluate products in order to obtain knowledge that can be applied when designing and making.

5. Wherever possible, design and technology activities will be planned within a school/class project, to add increased relevance.

6. Each design and technology activity will be planned on a design and technology unit of work planning sheet. Assessment opportunities will be identified at the planning stage.

7. Design and technology should always involve children in producing a product of quality.

8. The aim of our school is to develop a child's technological capability. By the end of Key Stage 2, the majority of children in our school should be able to:
   - work more independently and with confidence on design and technology activities;
   - draw upon a range of skills and techniques in order to inform their designing and making;
   - work confidently with a range of materials including those set down within the statutory Orders for Key Stages 1 and 2;
   - have developed a broad knowledge and understanding of the areas set down within the statutory Orders for Key Stages 1 and 2;
   - display a wide knowledge of how everyday objects are designed and made and provide suggestions as to why they are designed in a particular way and why particular materials are used to make them;
   - display enthusiasm for undertaking design and technology activities.

9. The teaching and learning of design and technology in our school should be both motivating and stimulating. Children should develop both a knowledge of the subject and an enthusiasm for undertaking further work in design and technology.

Design and technology in the National Curriculum

1. In the National Curriculum design and technology is set down under one Programme of Study called Design and Technology.

2. The main aim of design and technology is developing children's design and technology capability, through combining their designing and making skills with knowledge and understanding in order to design and make quality products.

3. At Key Stage 1 children are required to be provided with experience of working on design and technology activities having used a range of materials and components, including sheet materials, reclaimed materials, textiles, food and construction kits.

4. At Key Stage 2 children are required to be provided with experience of working on design and technology activities having used a range of materials and components, including stiff and flexible sheet materials, framework materials, textiles, food, electrical components, mechanical components and construction kits.

5. All children have an entitlement to access to the Programmes of Study matched to their knowledge, understanding and previous experience.

6. Coverage of design and technology is set down in the school's Programme for Design and Technology. The programme identifies the materials, designing skills, making skills and knowledge and understanding that will form the focus for each unit of work. Together the units of work for each Key Stage ensure total coverage of the statutory Orders.

7. The school's Programme for Design and Technology is attached to this policy as Appendix 1.

Teaching strategies and planning

1. It is important that the teacher identifies the most appropriate teaching strategy for a particular learning situation. The Scheme of Work provides guidance on the most effective methods for teaching specific areas of study.

2. Progression within design and technology is achieved by placing an ever increasing demand, within a wider range of materials, upon children to develop their designing and making skills which draw upon specific knowledge and understanding.

3. Progression will only be achieved with the whole school planning cooperatively and ensuring that teachers provide the planned focus for each unit of work taught.

In the classroom

1. Children are taught in their normal class group for design and technology.

2. All teachers are responsible for the teaching of design and technology.

3. Teachers should look for opportunities to praise cooperation and safe, considerate behaviour.

4. Children are encouraged to work as individuals, in pairs, in groups and also as a whole class when appropriate.

5. Classroom helpers will be provided with specific guidance on ways in which they are to work with children, the degree of independence that the children should be given and the specific aims and objectives for any activity that they are to oversee.

Equal opportunities and special needs

1. Activities both within and outside the classroom are planned in a way that encourages full and active participation by all children, irrespective of ability.

2. All units of work should show how the lessons are to be differentiated to cater for differing abilities.

3. Every effort will be made to ensure that activities are equally interesting to both boys and girls. Units of work will be planned to ensure that gender stereotyping is not reinforced.

Assessment and record keeping

1. Teachers will make brief notes on children's progress in design and technology at the end of each year. The notes should be kept to a minimum and yet provide enough information to inform the next teacher of the progress made and for use in the annual report to parents. The notes should be based on the Level Descriptions for design and technology.

   All assessments and records should comply fully with the school's Assessment Policy, copies of which are placed in teachers' policy folders.

2. All teachers will be responsible for ensuring that assessments are made at the end of each year so that updated records can be forwarded to the next teacher.

3. The school has a portfolio containing examples of children's work, matched to the Level Descriptions. All staff have discussed each piece of work and teachers' notes are attached explaining how the assessment was made. Due to the practical nature of design and technology, evidence of work undertaken by children can be in the form of teachers' notes or as a photographic record.

4. Teachers' own plans should indicate the focus for each unit of work and identify assessment opportunities.

Resources

1. All children should have opportunities to use information technology (IT), including word processors, databases, graphics programs and, at the upper end of the school, a computer control system.

2. Children will have opportunities to use the resources set out in the resources list.

3. Most design and technology equipment is kept within the central resources area. All equipment is readily accessible to the children. Children are given instructions in the safe and considerate use of resources, including care of consumables and materials that are not easy to store.

4. A full list of the resources available for design and technology has been set down by the design and technology coordinator. Curriculum equipment lists are attached to each policy as Appendix 2.

5. The design and technology coordinator is responsible for all design and technology resources, including ordering. The coordinator will ensure that the resource list is kept up to date.

Early years

1. Young children should be provided with as wide an experience as possible of working with materials including paper, card, construction kits, paints, glues, textiles, reclaimed materials and food. They should also be provided with opportunities to work with the associated tools and given direct instructions on the safe ways of working.

2. The reception teacher will plan appropriate activities in consultation with the year 1 teacher and use the school's Programme for Design and Technology as a structure for identifying appropriate activities.

Safety

1. Only older children should have access to tools such as circle cutters and glue guns and they must be under direct adult supervision. A separate area should be set aside for the use of glue guns and instructions should be given to the whole class on their safe use.

2. Sharp-pointed scissors can be used throughout the school but must be stored upright at all times.

3. Children should not use craft knives, which are for adult use only.

4. Direct safety instructions will be given to children each time they undertake a design and technology activity.

Working with food

1. An adult will be required to supervise activities involving cooking.

2. When undertaking food activities the appropriate health and safety procedures must be adhered to.

3. When working with food all children should be aware of the school rules regarding personal hygiene.

Before undertaking any food activity

1. Teachers should ensure that any helpers who are to participate in or lead any aspect of the food activities are made fully aware of all matters relating to health and safety when working with food and particularly those listed below.

2. Teachers should check the dietary needs of the children in their class to identify any foods that should not be available to specific children, or groups of children.

3. Any perishable food should be stored in the fridge.

4. Only the equipment in the food cupboard, which is for use with food only, should be used.

5. Glass and wooden items should never be used.

Setting out your food area

1. Ensure that the plastic sheets, specifically for use with food, cover the desk area.

2. Wipe down the sheets with a steriliser.

3. Ensure that all the utensils to be used have been cleaned and wiped with a steriliser.

4. Only use equipment set aside for food activities.

5. Set aside an area for children to wash their hands.

During food activities

1. Ensure that all children wear the aprons set aside for specific use during food activities.

2. Ensure that all children wash their hands.

3. Ensure that all children follow the school rules for working with food.

4. Teachers taking part in any food activity should dress appropriately and follow the same procedures as the children with regard to any rules regarding personal hygiene.

After the activity

1. Ensure that all equipment is cleaned and put away in the food cupboard.

2. Wipe down the work sheets with a steriliser.

Tasting food

1. Ensure that children use their own utensils when tasting food.

2. Red plastic spoons could be used for placing food on to plates and white spoons for tasting food.

   See the school's Health and Safety Manual for further guidance. For further information on any other aspect of health and safety relating to work with food consult the Local Authority guidance.

Review

This policy is reviewed by the staff and governors in the summer term. Parents are most welcome to request copies of this document and comments are invited from anyone involved in the life of the school.

3

A SCHEME OF WORK

Providing a Scheme of Work

It is expected that schools will plan to undertake one design and technology activity each term. To assist schools in planning their Scheme of Work for design and technology, this section provides a complete Scheme of Work for each Key Stage; six units of work for Key Stage 1 and twelve for Key Stage 2. Each unit of work provides clear guidance on what is to be taught and suitable tools and materials to use. Where relevant, each unit has been linked with other areas of the curriculum. This will assist schools in linking their design and technology projects to work already being undertaken in other subjects or to a school or class theme for the term. At the end of each unit of work, there are examples of the types of responses teachers might observe from the children, matched to different levels within Attainment Targets 1 and 2. This assessment guidance clarifies and extends the unit of work to give teachers greater support when planning and teaching that unit. It may also provide further ideas about skills, techniques or extension activities, as well as offering guidance on planning for differentiation.

The units of work   Although the units of work have been planned for either Key Stage 1 or 2, each can be adapted to suit any age within the primary range by taking the overall idea presented in the unit and setting down the details for the appropriate age group on a copy of the blank planning sheet (Appendix 1). Reference to the relevant Programmes of Study should be made throughout.

Planning a Scheme of Work   Schools need to decide how they will use the units of work provided.
- Choose the units that teachers find most interesting and adapt them to suit your own needs and add them to the existing units of work.
- Choose the units of work that teachers want to undertake and use the model provided for writing your own units to match existing plans.
- Choose the units of work that closely match the topics planned for each term. Where there is not a close match, use the format provided to write additional units of work.

To assist schools in writing or adapting the units of work provided, a blank planning sheet is supplied in Appendix 1, and clear guidance on writing to each heading can be found on pages 16–17.

A Scheme of Work for design and technology at Key Stage 1

There are six units of work for Key Stage 1 – one for each term. Each of the units has been mapped out on the following mapping sheet to show the coverage for each unit of work as well as the overall coverage of all six units of work throughout Key Stage 1. Teachers will need to decide which unit of work to teach each term and may need to adapt the units to suit the abilities and previous experience of the children. Teachers will need to use the blank mapping sheet provided in Appendix 2 in order to map out units that they have adapted or written from scratch.

KEY STAGE 1 MAPPING SHEET

Units of work	1. Cocktails	2. Souvenirs	3. Dips	4. Creating an environment	5. Puppets	6. A playground ride
Knowledge and understanding						
g. vocabulary	*		*			
f. health and safety	*		*			*
e. quality	*	*	*	*	*	*
c/d. products and applications	*	*	*			
b. structures				*		*
5a. mechanisms				*	*	*
Making skills						
f. evaluate their products as they are developed, identifying strengths and weaknesses			*			*
e. make suggestions about how to proceed	*	*	*	*	*	*
d. apply simple finishing techniques	*	*				
c. assemble, join and combine materials and components	*	*		*	*	*
b. measure, mark out, cut and shape	*	*	*	*	*	*
4a. select tools, materials and techniques	*	*	*	*	*	*
Designing skills						
f. consider their design ideas as these develop and identify strengths and weaknesses	*	*	*	*	*	*
e. make suggestions about how to proceed	*	*	*	*	*	*
d. develop and communicate design ideas through freehand drawing and modelling	*	*	*	*	*	*
c. develop their ideas through shaping, assembly and rearranging materials and components	*		*	*		
b. clarify their ideas through discussion	*	*	*	*	*	*
3a. draw on their own experience to help generate ideas		*	*	*	*	*
Materials						
construction kits				*		*
food	*		*			
textiles		*		*	*	
reclaimed materials		*		*		
2a. sheet materials		*		*	*	

COCKTAILS

Context
The children will examine a range of cocktail menus produced for them. They will look at an example of a cocktail that the teacher has made. The work can be linked to work on food and particularly fruits. The project can also be linked to a party or theme that forms the term's topic. If it is coming up to Christmas time then the children can make cocktails for their party.

Outcome
Making a cocktail from soft drinks and shaping fruit in order to decorate their drinks.

Health and Safety

Before undertaking any work with food ensure that:
■ You are fully aware of any dietary requirements that may influence the foods that you can make available to the children.
■ You are fully aware of any health and safety requirements set down within the school's Design and Technology and Health and Safety Policies.

When undertaking work with food:
■ Ensure that a specific food area is established and that all tools and clothing used are for use within food activities only.
■ Ensure that all children are fully aware of any rules regarding personal hygiene.

For further guidance:
Refer to the sample Design and Technology Policy, pages 26–30, which provides specific guidance on health and safety matters relating to the teaching of food technology.

Cross-curricular links
HEALTH EDUCATION & SCIENCE
The work can support a health-related project that looks at healthy foods and healthy eating.

RESOURCES FOR INVESTIGATING
■ A selection of commercially-produced cocktail menus for children.
■ A selection of utensils for cutting and shaping fruit.
■ A selection of soft drinks, including fresh fruit drinks and the equivalent cordial mix. Mixer drinks including water and lemonade.

RESOURCES FOR MAKING
■ fresh fruit drinks
■ cordials
■ lemonade
■ food colourings
■ sugar
■ selection of tools for shaping the fruit

■ water
■ selection of fresh fruits
■ white paper napkins
■ spoons

ACTIVITIES

Focused practical task

■ Provide opportunities for the children to experiment with the drinks available to create different colours. Set the task of creating a number of different colours and ask the children to record what they have done so that they can recreate them if they decide to use them in their cocktails.

- Encourage the children to practise using the available utensils to create different shapes and figures from the fruit.
- Show the children how to use the food colourings.
- Take every opportunity to talk about drinks that are healthy and those that are less healthy.
- Ask the children to compare the drinks they mix and to say which they think are the healthiest and why.
- Provide opportunities for children to taste their mixes. Explain that they should take only a small taste.
- Encourage the children to explore the different effects and tastes that can be created by adding lemonade or water to their drinks.

Investigative, disassembly and evaluative activities

- Give the children the cocktail menus and ask them to say which cocktails sound and look the best. Ask them to give reasons for their choice.
- Ask the children to identify ideas that they might use in making their cocktails and to explain how they would set about this.
- Make a cocktail of your own and give it an appropriate name. Ask the children to say why they think you gave it that name.
- Ask the children to write down how they would set about making a cocktail like the one you have made. Ask them to explain to others how they would make it.
- Provide an opportunity for the children to attempt to make a cocktail like yours. What does it taste like? If they don't like the taste, can they think of a way of improving it?

Design and make assignment

- Explain to the children that they are going to make a cocktail. You may decide on a theme or let them choose their own. Your decision will depend upon what they have investigated and evaluated beforehand.
- Tell the children that they should each choose a suitable name for their cocktail and that the drink should reflect the name they give it.
- Encourage the children to make drawings, sketches or paintings to indicate what their cocktails might look like.
- The children may wish to make drawings of their cocktails using a graphics program on the computer.
- Remind the children of the factors that make a drink healthy.
- Ask the children to consider how they would make their drink look attractive so that customers would want to buy it.
- Remind the children that, as good as their drink might look, nobody is going to buy it if it doesn't taste nice.
- Ask the children to think about the order in which they are going to make their cocktail – for instance, if they make the fruit shape first, it may discolour whilst they are making the drink.
- The children should be provided with an opportunity to display their drinks and to evaluate them by tasting. Other children can evaluate how good the drinks look and whether they would buy one. Encourage them to give reasons for their responses.

EXTENSION ACTIVITIES

- A place mat to present the cocktail on can be decorated appropriately.
- Menus can be designed for the cocktails.

ASSESSMENT   For assessment purposes, teachers might observe the following types of responses as children work through this unit.

at Level 1   Designing
- They generate ideas by mixing the drinks to create a desired colour.
- They generate ideas by cutting the fruit into various shapes.
- They recognise that some fruits are easier to model than others.
- They make simple drawings showing the colour of their cocktails.

Making
- The children explain what they are making and what they are going to use.
- They select peaches, hollow them out and remove the stones to create boats.

at Level 2   Designing
- They mix the drinks to create the colour required for their cocktails.
- They create drawings of their cocktails using Flare on the Archimedes.
- They make their cocktails and explain that they do not taste very nice because they added too much lemon. If they made more cocktails, they would use orange instead of lemon.

Making
- They select from a range of appropriate tools to cut and shape the fruit.
- They mix two drinks to create the colour required and then add lemonade because they want the bubble effect so it looks 'as if you are looking into a fish tank'.
- They explain that their drinks look like a fish tank but that the fish they have made from the fruit are not floating in the middle as they wanted.

at Level 3   Designing
- They make several designs, giving each a name matched to its colour and/or ingredients. They explain that the best looking cocktail is far too sweet and is therefore unhealthy, so they are making the next best.
- They explain how they can create an umbrella to place on a cocktail to make it look more attractive. They explain that they had thought of other ideas, including a boat to float on the top of the drink and of shaping half a fruit to sit completely over the top of the drink with a hole in the middle for the straw.
- They design a drink with one colour at the top and another at the bottom which they know they can make because they have discovered that some drinks don't mix unless you stir them.
- They clearly label their sketches.

Making
- They create the drink but do not add the lemonade because they decide that, by the time they have made the umbrella for the top, the drink will have gone flat.
- They trim the rim of the glass with small pieces of fruit and place the cocktail on a matching coloured mat because they say it looks better.
- The cocktail looks similar to their original sketch but they explain that they have changed the fruit they placed in the drink because the first type they used sank and they wanted it to float on the surface.

SOUVENIRS

Context As part of a school visit children can take the opportunity to look round a souvenir shop. The teacher can bring to their attention the range of souvenirs but particularly ones that present the children with ideas for designing and making with the range of materials they have available. They could look at bookmarks or clay pots that could be used as the basis for developing their own design ideas. The teacher should return to school with a number of souvenirs that can be used to stimulate children's thinking.

Outcome Making a souvenir for an area or place that they have visited.

Cross-curricular links
GEOGRAPHY The work helps to develop children's understanding of a contrasting locality.

RESOURCES FOR INVESTIGATING
- A collection of souvenirs purchased from a place visited on a school trip. The collection should include a number of souvenirs that the children can make for themselves – e.g. a bookmark.

RESOURCES FOR MAKING
- This will depend upon the types of souvenirs that you have collected and the materials that you have decided will form the focus for this unit of work. Choose a collection of souvenirs that will focus the children's attention upon certain materials – e.g. textiles. This is a particularly good unit for working with textiles. Children may decide to make bags, bookmarks, belts or badges.

ACTIVITIES

Focused practical task

- Provide opportunities for the children to learn how to work with textiles. They should be able to join through sewing and sticking and produce simple sewing patterns.
- Teach the children how to decorate their textiles by sticking and sewing other materials such as buttons, sequins etc.
- Demonstrate how to sew two pieces of material together and to stuff the insides to produce a soft toy effect.
- Demonstrate and teach the skills and tools involved in making the souvenirs that you have collected.

Investigative, disassembly and evaluative activities

- Make a display of the souvenirs and other items brought back from the school trip. It is a good idea to include photographs in the display, as these will provide children with ideas.
- Draw the children's attention to the different souvenirs and talk about the materials from which they are made.
- Ask the children how they know that these are souvenirs.
- On a sheet of drawing paper, ask the children to draw one of the souvenirs and around it write or draw everything they know about it. They can, for

instance, say where it is from, what it is made from, what they like about it. Some children will need to be asked questions to encourage them to think carefully about their souvenir.

Design and make assignment

- Ask the children to set down their ideas for a souvenir that they might make for a visit to a specific place. Depending upon their age and ability, either you or they can choose the place.
- Recap on the souvenirs on display and how they might be used to generate ideas.
- Show the children the materials and tools that are available. Explain that there are no other materials available and so they must plan using those on display.
- Ask children to explain what they are going to make, how they are to set about making it and the order in which they are to tackle each part. The children may present this as a series of drawings and text, a labelled diagram or a verbal explanation.
- Once you are confident that they know what they are aiming to achieve, they can begin making their souvenir. Every so often, stop a group and ask them to explain to others what they have done and to say if they have changed anything from their original ideas.
- Emphasise throughout the need for quality in what they make. If you do not feel that something a child is making is of the quality that could be achieved, ask how it might be improved and whether somebody would want to buy it.
- Encourage the children to look back at their original ideas to compare them with what they are making.
- Once the souvenirs are completed, ask the children to evaluate their own work. Encourage them to evaluate other children's souvenirs in a supportive manner.
- Ask the children whether they could improve their souvenirs.

EXTENSION ACTIVITIES

- The children can make boxes for their souvenirs.
- They can produce advertisements for their souvenirs.

ASSESSMENT For assessment purposes, teachers might observe the following types of responses as children work through this unit.

at Level 1 **Designing**
- They relate the leather bookmarks they bought on the school trip to their own ideas.
- They draw a picture of the bookmark and draw their pattern on it. They explain the drawing.

Making
- They explain what they are making and that they are using thick card to make it from.
- They select suitable thicknesses of card and use scissors to cut an appropriate length matched against a book. They cut along the base of the card bookmarks to create the same effect as on the leather bookmarks they bought on the school trip.

at Level 2 **Designing**
- They explain how they will make souvenirs from clay, drawing upon skills learnt in art lessons. They explain how they can use sausage-shaped lengths of clay coiled around to make bowls like those seen in the souvenir shop.
- They make simple drawings to explain what they are going to do.
- When they have finished, they explain that they should have made the clay smoother, because their pot was too rough down one side.

Making

- They choose appropriate tools for shaping and smoothing their clay pots.
- They use the shaping tools to create a design on the outside of their pots.
- They say that people would like their pots because they are in bright colours.

at Level 3 **Designing**

- They drew upon ideas from the souvenir shop where they had seen the purses.
- They design a purse and, when they talk through their ideas with their teacher, they explain the ideas that they have rejected.
- They produce labelled sketches showing how the purses are to be made and what the finished product might look like.
- When they design their felt purses, they draw complicated straps for them. They eventually decide to make the purses without straps because the straps will take too long to make. They are clearly aware that time is a factor that has to be taken into consideration at the design stage.

Making

- They gather the equipment and materials together to cut out two pieces of felt and to sew them together. This clearly indicates that the children have planned in advance what they were going to do and what they would require.
- They use the needle and thread to sew two pieces of felt together. They know a range of sewing styles and use the knowledge to decorate the purse.
- They cut out the felt shapes accurately and pin them together prior to sewing.
- Their finished purses look similar to their drawings and they explain that they have not included all the decorations because they felt that the purses didn't look right with too many decorations.

DIPS

Context   The dips can be made as part of a theme party that the class are planning. This unit can be linked with Unit 1: Cocktails, with the children working on producing various drinks and dips for an end of school party. The work can be linked to work on food and particularly vegetables.

Outcome   Making a dip and appropriately shaped vegetables to accompany it.

Cross-curricular links
HEALTH EDUCATION
& SCIENCE   The unit can be linked to work on healthy eating.

Health and Safety   **Before undertaking any work with food ensure that:**
- You are fully aware of any dietary requirements that may influence the foods that you can make available to the children.
- You are fully aware of any health and safety requirements set down within the school's Design and Technology and Health and Safety Policies.

When undertaking work with food:
- Ensure that a specific food area is established and that all tools and clothing used are for use within food activities only.
- Ensure that all children are fully aware of any rules regarding personal hygiene.

For further guidance:
Refer to the sample Design and Technology Policy, pages 26–30, which provides specific guidance on health and safety matters relating to the teaching of food technology.

RESOURCES FOR INVESTIGATING
- As below, plus any photographs of party food showing dips. Cookery books can be displayed to provide additional ideas.

RESOURCES FOR MAKING
- attractive plates and bowls (not glass)
- selection of vegetables or fruits
- selection of crisps
- mixing bowl
- carton of full fat or low fat soft cheese
- milk
- tablespoon
- wooden spoon

ACTIVITIES   Focused practical task

- Provide a range of utensils for cutting and shaping and several fruits and vegetables.
- Show the children how to set up a safe and hygienic area in which to work.
- Demonstrate how to use each of the implements safely and correctly.
- Ask the children to shape the selection of foods into slices, pieces, squares or circles, using the tools provided. Older children can be challenged to make interesting figures which combine different shapes.

Investigative, disassembly and evaluative activities

- Make a collection of foods and explain to the children which foods are healthier and why. Have a mixture of foods that you can sort into groups of healthy and less healthy. Ensure that the children can clearly see that fruit and vegetables are healthy foods.
- Make up a dip and explain that they are often made for parties. Explain each stage, especially the dip that you make, after telling the children that they are going to make one of their own for the class. Allow the children to select items to put into the dip and to taste the dip you have made. Children should be encouraged to say which taste they prefer and which of the dipping items they consider to be more healthy and why.
- Use the correct vocabulary for all the utensils throughout.

Design and make assignment

- Collect several food items that children can use for dipping. Ask them to consider which they consider to be healthy and which they prefer.
- Remind the children how to make the dip and recap on how they should set up their working area.
- Remind the children to wash their hands before preparing food.
- Ask the children to make a list of the foods they are going to use. Then ask them to make a drawing showing what the dipping items might look like once they are made.
- Once you are sure the children have a good idea of what they are going to make, allow them to begin making.
- Remind the children throughout of the correct way of using each utensil. Emphasise the need for the dip and dipping items to look nice if other children are going to want to eat them.
- Throughout, encourage the children to use the correct terms for the utensils and foods when asking for items or explaining what they are doing.

EXTENSION ACTIVITY

- The children can design and make a mat on which to present the bowl of dip. They should consider the colours of the foods and utensils that they have used when designing their mats.

ASSESSMENT   For assessment purposes, teachers might observe the following types of responses as children work through this unit.

at Level 1   **Designing**
- They generate ideas for items to dip when they are cutting and shaping the foods.
- They decide to use long strips of carrots as they can be placed in the dip easily.
- They make a simple drawing of their dipping items.

Making
- They explain the foods they are using to make the dipping items.
- They select carrots and cucumber from the selection of foods and shape them.

at Level 2   **Designing**
- The children make drawings of what their dipping items will look like and can explain how they will create each of the shapes.
- They find a photograph in a children's cookery book and explain that they are going to shape their carrots with the crinkler like the ones in the book.
- They look at their finished dip and decide to present the dip in a clean bowl with the items for dipping arranged attractively around it on a plate. They think this will make people want to eat it.

Making
- They select healthy foods from a range and explain that they have chosen the foods that are good for you.
- They correctly use a range of utensils to make the fruits and vegetables into different shapes.

at Level 3 **Designing**
- They make several designs and discuss which they think is the best. The design they decide to make is not the one they would most like to eat because they have had to choose foods that can be shaped into long strips for dipping.
- They explain to the teacher how they will make their dip.
- They use appropriate vocabulary when referring to the cooking utensils.
- They make and label sketches of their dipping items.

Making
- They gather their equipment and food items together to make their dip first and then clear away and prepare to make the dipping items.
- They use the utensils correctly.
- They arrange the foods for dipping symmetrically around the dip.
- They peel and slice the carrots before using the crinkler to make more interesting shapes. They cut off the tops of the tomatoes and fill them with some of the dip.
- Their dips and items for dipping look similar to their designs.
- They evaluate their dip, they explain that it tastes as good as they thought it would but that they should have scooped the insides out of the tomatoes before adding the dip because, when they ate it, it squirted all over them.

CREATING AN ENVIRONMENT

Context A role-play area is a common sight in early year classes. Opportunities can be taken to develop the area to make links with design and technology. The area can be changed at regular intervals to link in with the term's project or to provide opportunities for young children to experience playing in a different environment. In some schools young children may visit a local shop such as a hairdressers and return to school to establish a similar environment in their own classroom.

Children can consider the main aspects of the environment they have visited and make some of the objects they have seen.

Outcome Creating an environment within the classroom or other area of the school which presents children with opportunities for role play.

Making objects for the role-play area.
Examples of environments for the role-play area include:
– a hairdressers   – a garage
– a post office   – a spaceship
– a supermarket   – a kitchen
– a local shop

Cross-curricular links
SCIENCE Sorting and classifying materials and matching the properties of the materials to their function.

RESOURCES FOR INVESTIGATING
■ A visit to a local shop, industry, museum or farm.
■ A display based on a project that the class are undertaking on the past or a recent event.

RESOURCES FOR MAKING
■ A wide selection of reclaimed materials – see pages 103–104 for guidance.

ACTIVITIES   Focused practical task

■ Provide the children with opportunities to join, stick and paint a variety of objects, using reclaimed materials.
■ Allow children the opportunity to choose their own items to make, to play with the materials and to cut and join in the ways that they choose. Introduce the children to ways of preparing their own working areas and of cutting, sticking and joining.
■ Use the opportunity to sort and classify the reclaimed materials for storage in a way that develops children's vocabulary – e.g. round, tubes, clear, transparent.

Investigative, disassembly and evaluative activities

- Decide upon the environment that you are going to create and explain to the children what they are going to do and where they will set up their environment. You may decide to make a start by placing some objects in the area you have set aside, to give the children a feel for the task ahead.
- Take the children on a visit or study a film or books of an event in history to provide a focus for their ideas. Ask the children to explain how they know what environment they are in. Encourage them to identify the main features that characterise that environment.
- Ask the children to observe how people behave in the environment they are to create. They will be given opportunities to play within the environment that they create.
- Collect and sort as much information as you can from the environment that you visited. Record the visit by taking photographs and displaying them in the classroom.
- Encourage the children to use the correct terms when describing the people and objects within the environment.

Design and make assignment

- Ask the children to tell you what they think should go into the environment they are to create. Make a list and add to it anything they have missed out. Ask them to consider where the items will go. For example, if you were creating a hairdresser's salon, it is likely that there would be a till, a telephone and a book for making appointments placed near the entrance.
- Show the children the materials that they have to work with and remind them how they should set out their working areas.
- Ask the children to make a drawing of what they want to make. Talk through their ideas with less able or younger children. When you are sure that they know what they are going to make and how, let them begin.
- Remind the children throughout of the correct ways of working and ensure that they use the tools and other equipment correctly and safely.
- Once the environment has been created, allow time for the children to play in it. Gather the children together and ask them if they are pleased with the environment that they have created. Is there anything missing or that they saw on their visit but were unable to recreate?

EXTENSION ACTIVITY

- The children can invite people from the place that they visited to come along and see their environment and to evaluate it. This will help them to develop some understanding of the value of other people's views.

ASSESSMENT   For assessment purposes, teachers might observe the following types of responses as children work through this unit.

at Level 1   **Designing**
- They create ideas by placing different shaped boxes alongside and on top of each other.
- They recognise the shape and size of the object that they are going to make.
- They talk to the teacher about what they are going to make. They can place several boxes together to give a guide to what shape and size their object is going to be.

Making
- While they are making, they explain to the teacher what they are doing and the materials they are using.
- They join several boxes to create the shape they require.

at Level 2

Designing
- Their design clearly shows that they understand that they can make a slit in the box and slide a clear piece of plastic through to create a window effect.
- They have made a simple drawing to show what they are going to make.
- They explain that they should have taken more care when painting their model as the paint has dripped all down the side.

Making
- They select appropriate shapes and sizes from the reclaimed materials available.
- They slide boxes together to make a longer shape.
- They explain that their object is going to be bigger than the real thing, seen on their visit.

at Level 3

Designing
- When designing, they recognise that there are some items that they cannot make because they don't have materials of the right size, shape or texture.
- They set down a design and explain it to the teacher. In discussion, they explain their other ideas.
- They place a heavy weight in the bottom of their finished object because it was so top heavy that it would have fallen over otherwise.
- They produce a labelled sketch showing the materials that will be used to make the object.

Making
- They are clearly thinking ahead when working and use tools with care and confidence.
- Their final object clearly shows that they have used a range of techniques to cut and join the various materials used.
- Their object looks similar to their design.
- They explain the changes they have made to their original design.

<table>
<tr><td>KEY STAGE 1
UNIT OF WORK
5</td><td></td></tr>
</table>

PUPPETS

Context Children may visit a puppet show or watch a performance in their own school. They may make simple puppets based around a story or the class theme. The teacher can set up a puppet theatre area so that the children can use their puppets to act out a simple story they know.

Outcome Making a simple mitten or stick puppet.

Cross-curricular links
ENGLISH Speaking and listening occur when the children use their puppets to perform a play.

RESOURCES FOR INVESTIGATING
- A visit by a puppet making group.
- A mitten and a stick puppet made by the teacher.

RESOURCES FOR MAKING
- Textiles for making mitten puppets. The mitten puppet is easier to sew together than a glove puppet for young children.
- Sewing equipment including needles and thread.
- A wide range of decorative materials suitable for sticking or sewing on to the mittens.
- Sticks for sticking or taping to the card puppets.

ACTIVITIES

Focused practical task

- Provide opportunities for the children to stick and stitch materials on to felt and card.
- Explain how to set out the working area and point out the correct ways of using the available needles, thread and glues.

Investigative, disassembly and evaluative activities

- Make an example of a mitten puppet and a stick puppet for the children to study. Ask the children to describe what the puppets are made from. Can they explain how they would make a similar puppet?
- Provide older children with art paper to make drawings of the puppets and encourage them to label each of the parts, explaining the colour and type of material used. Ask them to write down how they think the puppet was made.
- If possible, invite a puppet-making group to visit the school. They can explain how to make simple puppets and put on a production that will provide even more ideas for the children to use and adapt.

MAKING A MITTEN PUPPET

Make a paper pattern to the correct size. Cut out two mitten-shaped pieces of crimplene or cotton type material, a non-fraying material, which is easier to sew. Stitch the two pieces with right sides together and then turn inside out. Decorations can now be stuck or sewn on to the mitten puppet to create the character.

MAKING A STICK PUPPET

Cut out a figure or face from card and attach suitable pieces of textiles and other decorations to create the character. Once complete, use tape to attach the figure to a suitable length of stick or dowel.

Design and make assignment

- Show the children the range of available materials. Remind them of the two methods they have been shown for making puppets.
- Ask the children to decide which method they are going to use to make their puppets. They can draw their puppet, showing the materials they are going to use to create the character. Children should plan to use the available materials.
- Talk through the children's ideas and, once they are clear about how they are going to set about making their puppets, they can begin making.
- At intervals, ask the children to explain what they have done. Encourage them to consult their designs to see if they match what they are making.
- When the puppets are complete, ask the children to explain what they think of them, consider whether they look similar to their designs and suggest any ways in which they might improve their puppets. Can they, for instance, make them look more fierce or sad or attractive?

EXTENSION ACTIVITY

- Set up a simple stage in the classroom and allow the children to perform a short play, using the puppets they have made.

ASSESSMENT   For assessment purposes, teachers might observe the following types of responses as children work through this unit.

at Level 1   **Designing**
- They place their decorative materials on to their mitten puppets when deciding where best to stick them.
- They choose suitable sequins to create eyes for their puppets.
- They explain what they are going to make and how they are going to make it.

Making
- They explain that they are using felt and buttons to decorate their stick puppet.
- They select suitable decorations to stick on their puppets from a narrow range available.

at Level 2   **Designing**
- Their designs show an understanding that the card figure is cut out first and then materials and finally the decorations are added.
- They create a simple drawing to help communicate their ideas.
- They decide to change the felt eyes for sequins because they think they look more like eyes.

Making
- They select suitable materials from a range of decorative materials to hand.
- They explain why they chose particular materials at each stage of making.
- They explain that they like the ears that stick out because they will flop about when they move the stick puppet up and down.

at Level 3 **Designing**

- They explain that they decided to make a stick puppet because this allowed them to make a bigger, full-length figure.
- They attach two legs with split pins so that the legs swing about. They explain that the puppet makers who visited school had used the same idea.
- Their designs are clearly labelled to show the materials to be used.

Making

- They thought out the order in which they would make each part of the puppet. Tools and equipment are gathered together and the work area is prepared before they begin work.
- They pin down the paper pattern and cut out the mitten shapes. They pin them together before sewing.
- They use a stitching pattern to produce a pattern around the eyes of their character.
- Their mitten puppet looks like their original design except for the mouth which they replaced with a different coloured material because the original did not show up clearly against the body colour they had chosen.

A PLAYGROUND RIDE

Context   The children may have visited a local playground to observe and play on the various rides. They can make sketches of the rides and take photographs that will form the basis of a class display. They will have studied the rides at firsthand, observing how they work and relating their movement to ways in which they can be created with materials back in the classroom.

Outcome   Making a small-scale playground ride.

Cross-curricular links
GEOGRAPHY   The work could link with a study of the local environment, if the school has a playground in the locality, or could form part of work that develops from a school trip.

RESOURCES FOR INVESTIGATING
- A visit to a playground or a play area.

RESOURCES FOR MAKING
- A wide range of reclaimed materials – see pages 103–104 for guidance.
- A selection of construction kits suitable for making models of moving and static playground rides, such as Lego and First Gear.

ACTIVITIES

Focused practical task

- Provide opportunities for children to work with the construction kits. Encourage them to explore all the parts so that they are aware of the types of movement or structure that can be created.
- Display several of the structures that the children make, so that other children can look at them and investigate how they were made.
- If there are activity cards with any of your construction kits, pick out those that focus upon movements and structures that you feel children may be able to incorporate into their future designs. Ask the children to recreate the constructions shown on the activity cards you have chosen.

Investigative, disassembly and evaluative activities

- Visit a playground and observe the different rides. Draw to the children's attention the different structures for climbing over or through. Look closely at the moving rides and describe them to the children. Use the appropriate vocabulary when describing structures or forms of movement.
- Relate the structures and movements to any that have been created in the classroom using the construction kits.
- Take photographs of the rides so that the children can use them to prompt ideas back in the classroom.
- Provide children with materials for sketching and ask them to sketch one of the rides. Older children can label the rides and show in more detail how some parts work.

Design and make assignment

- Set out an area where your playground is going to be set up – it may be along one side of the classroom if space is at a premium.
- Make a display of the photographs and sketches you made when you visited the playground. Draw to the children's attention all the things they saw.
- Show the children all the reclaimed materials that are available for making their rides. Remind them of any techniques you may have taught them for joining materials together. Ask the children to suggest ways in which any of the reclaimed materials might be used. They may suggest that the string can be used for climbing or that the wood strips can be used to make a simple climbing frame.
- Draw to the children's attention the construction kits that are available. Ask them to consider ways in which the kits might be used by looking at some of the models they have created.
- Tell the children that they can make their rides from reclaimed materials, construction kits or both.
- Ask the children to make a drawing of a playground ride that they would like to make. Older or more able children can be asked to come up with several ideas and then to choose the best one to make. Remind the children that they should only design a playground ride that can be made from the materials that you have shown them.
- Talk through the children's ideas and, once you feel that they are clear about what they are going to make, they can set up their working area and select their materials and tools.
- At various stages, ask the children to talk through what they have made and explain what they are going to do next. They should be encouraged to evaluate the rides as they are making them, looking for any improvements that could be made.
- Once the playground rides are made, ask the children to explain how they made them and to evaluate how suitable their rides would be if they were to be made and placed in a real playground.

EXTENSION ACTIVITY

- Children may have seen rides surrounded by materials that cushion any child falling off. If children wish to ensure that their ride is safe, then they can place it in a suitably-sized shallow cardboard box and consider the types of materials that they could place around the ride that would cushion a fall. Children should be encouraged to investigate a range of ideas for suitable materials, e.g. cutting up paper, crumpling up paper, grating thick card. They can place several different materials into a box and ask others to evaluate which is the best in terms of its cushioning effect and whether it will be blown or kicked away easily, leaving no protection.

ASSESSMENT For assessment purposes, teachers might observe the following types of responses as children work through this unit.

at Level 1 **Designing**
- They make a collection of reclaimed materials and place them together to create the correct shape for their ride.
- They use aluminium foil for a slide because they say things slide down it easily.
- They explain what they are going to make.

Making
- They explain that they are cutting the sides out of a box and then hanging string from it to make a rope climb.
- They cut several materials and stick them together.

at Level 2   **Designing**
- They make a climbing frame from wood strips because they have used them before and know they make strong frames.
- They make a drawing to show what they are making. They use one of the photographs taken on the visit to the playground to give them ideas.
- They evaluate their ideas for making and explain that First Gear would make a better roundabout than the one they have designed using a circular box.

Making
- They choose suitable materials from the range provided. They are familiar with the range of ways in which several of the construction kits can be joined to create structures and movements.
- Their playground ride shows an ability to join materials in more than one way.
- As they are working they make a judgment about how their work will turn out.

at Level 3   **Designing**
- They make several designs but recognise that one of their designs will take too long to make and that another will use too much of the construction kits available.
- They explain how they would make what they have designed and, when asked, explain other ideas that they had but had not designed.
- They explain that they need to create a triangular shape for their swing otherwise it will fall down.
- They make labelled sketches clearly showing the materials to be used to make their ride.

Making
- They gather together the materials and tools they require before they begin making. They choose appropriate techniques for making.
- As they are making, they realise that their triangular structure requires reinforcement with cross pieces.
- They cut wood strips and demonstrate skill at assembling and disassembling particular construction kits.
- Their rides look similar to their designs and they explain where changes have been made.

A Scheme of Work for design and technology at Key Stage 2

There are twelve units of work for Key Stage 2 – one for each term. Each of the units has been mapped out on the following mapping sheet to show the coverage for each unit of work as well as the overall coverage of all twelve units of work throughout Key Stage 2. Teachers will need to decide which unit of work to teach each term and may need to adapt the units to suit the abilities and previous experience of the children. Teachers will need to use the blank mapping sheet provided in Appendix 3 in order to map out units that they have adapted or written from scratch.

KEY STAGE 2 MAPPING SHEET (1)

	Units of work	1. Greeting cards	2. Blackpool Illuminations	3. Transport	4. Activity centres	5. Yoghurts	6. Packaging
Materials	2a. stiff and flexible sheet materials	*	*		*		*
	framework materials		*	*	*		
	mouldable materials						
	textiles						
	food					*	
	electrical components		*	*	*		
	mechanical components	*		*	*		
	construction kits		*		*		
Designing skills	3a. use information sources in their designing	*	*	*	*	*	*
	b. generate ideas, considering the users and purposes						
	c. clarify ideas, develop criteria and suggest ways forward						
	d. consider appearance, function, reliability						
	e. communicate and model ideas			*	*	*	*
	f. develop a planned sequence and suggest alternatives		*		*	*	*
	g. evaluate design ideas against user and purpose and suggest ways forward	*		*	*		*
Making skills	4a. select appropriate tools, materials and techniques	*	*	*	*		*
	b. measure, mark out, cut and shape		*	*	*		*
	c. join and combine materials and components		*		*	*	*
	d. apply additional finishing techniques	*	*			*	
	e. select and plan use of materials, equipment and processes	*		*	*	*	
	f. evaluate and test their product		*	*	*	*	*
	g. implement improvements they have identified			*			*
Knowledge and understanding	5a. materials and components – working characteristics relate to their use						
	b. materials and components – combining and mixing to make more useful					*	
	c. control – mechanical	*		*	*		
	d. control – electrical		*	*	*		
	e. structures		*	*			
	f./g. products and applications	*		*	*	*	*
	h./i. quality	*	*	*	*	*	*
	j. health and safety					*	
	k. vocabulary	*	*	*	*		

KEY STAGE 2 MAPPING SHEET (2)

Units of work	7. Making a meal of it	8. Torches	9. A carrying bag	10. Reading books	11. Soap dishes	12. Lifting and lowering
Knowledge and understanding						
k. vocabulary	*	*				*
j. health and safety	*				*	
h./i. quality	*	*	*	*	*	*
f./g. products and applications	*	*	*	*	*	
e. structures						*
d. control – electrical		*				
c. control – mechanical		*				*
b. materials and components – combining and mixing to make more useful	*		*			*
5a. materials and components – working characteristics relate to their use		*	*		*	
Making skills						
g. implement improvements they have identified			*			
f. evaluate and test their product	*	*	*	*	*	*
e. select and plan use of materials, equipment and processes		*	*	*		*
d. apply additional finishing techniques			*			
c. join and combine materials and components		*	*			*
b. measure, mark out, cut and shape			*	*	*	
4a. select appropriate tools, materials and techniques	*	*	*	*	*	*
Designing skills						
g. evaluate design ideas against user and purpose and suggest ways forward			*	*	*	
f. develop a planned sequence and suggest alternatives	*		*	*	*	*
e. communicate and model ideas		*	*			*
d. consider appearance, function, reliability						
c. clarify ideas, develop criteria and suggest ways forward	*	*	*	*	*	*
b. generate ideas, considering the users and purposes						
3. use information sources in their designing						
Materials						
construction kits						*
mechanical components		*		*		*
electrical components		*				
food	*					
textiles			*			
mouldable materials		*			*	
framework materials						*
2a. stiff and flexible sheet materials				*		*

GREETING CARDS

Context
This unit of work will be based around a specific celebration which may be a birthday, party or religious festival. Children will study a number of greeting cards to gather information regarding layout, colour and text. They will link the information on the card to their knowledge of the celebration.

Outcome
Making a greeting card for a specific celebration.

Cross-curricular links
INFORMATION TECHNOLOGY
Using a word processor and graphics program.

ART
Providing opportunities to apply skills learnt in art within a design and technology context.

RESOURCES FOR INVESTIGATING
- A collection of greeting cards that provide a stimulus for the children. The collection of cards is important in setting the level at which the children will design and make. For younger children include cards that have no moving parts and provide ideas for colours, shapes and layout of text. For older children choose cards that are more complex and include moving parts or novel pop-up sections.

RESOURCES FOR MAKING
- Plentiful supplies of paper for modelling.
- A range of different coloured cards.
- Graphic materials for drawing and colouring.
- A computer system complete with graphics program, word processor and colour printer.

ACTIVITIES

Focused practical task

- Provide a number of cards with moving parts. Reproduce one of the moving parts from paper and card and explain to the children how you made it. Draw to their attention any specific techniques or materials that you have used, such as using paper clips to form linkages or reinforcing the paper with strips of card where a sliding mechanism was required.
- Using some pre-made models, explain ways in which card can be shaped to produce interesting effects, such as the 'spring' or 'concertina' by folding. Show them how to mark and fold pieces of card with accuracy.
- Provide opportunities for the children to practise marking and scoring to create specific shapes or effects from card.
- Demonstrate to the children how to use the word processor to print out messages to include in their cards. The children should be able to:
  - load the word processor
  - enter the text
  - position the text
  - change the font and size of the text
  - add a border or picture from art drawings available
  - print out their text.

- Teach the children how to use an art package such as Flare to make drawings and designs for their cards. Children should be able to:
  - load the program
  - create shapes and designs
  - be familiar with the main tools available for creating pictures
  - change colours to create the colour required
  - save a picture
  - print out their pictures.
- Encourage the children to explore ways of making or creating text required by painting or making their own stencils.

Investigative, disassembly and evaluative activities

- Provide a wide range of greeting cards for children to observe. Talk through each of the cards with the children and draw their attention to:
  - the colours associated with the celebration or the style of art used
  - the layout of the greeting text
  - the messages inside the cards
  - any moving parts or novel ideas.
- Provide opportunities to look closely at the cards. Ask the children to make a note of anything of interest or that might be included in their own designs. Provide time for the children to explain to the class the things that they have noted.
- Ask the children to open a card and observe:
  - how it stands up
  - places where it might have been strengthened.
- Ask the children to look closely at any moving parts or pop-up sections and to try and work out how the designer created the movement.
- The children could take apart some of the cards if it helps them to see more clearly how the cards were made.

Design and make assignment

- Make a list of all the ideas the children have for making a greeting card. Discuss whether all their ideas are suitable for the card they are going to make.
- Explain that they are going to make a card and that they should try to achieve the same quality of card as the ones they have been looking at. They will need to take great care at each stage of making.
- Explain that they cannot possibly begin to make their cards until they have thought through what they are going to make and the order in which they are going to make each part. They will need to draw their designs with care, label each part and explain the order in which they are going to tackle making the parts of the card. The children should be told that if their designs are good, somebody else should be able to follow them to make the card themselves.
- Once the designs are complete, talk them through with the children to ensure that they are clear about what they are setting out to achieve.
- Explain the need to use the expensive card with care. Show the children how they should model any moving part with paper before attempting to make it from the card. Point out that card is less flexible to bend and shape than the paper they are to use for modelling.
- Encourage the children to evaluate their own work and that of others throughout the process of making. Continue to emphasise the need to maintain the quality of the final product.

EXTENSION ACTIVITY

- Those children who finish their work before the others can set about designing and making an envelope for their cards. Allow the children time to evaluate your collection of envelopes before making their own. Include envelopes that have drawings on them that match the designs of the cards as this will present a greater challenge to the children and a corporate design feature.

ASSESSMENT

For assessment purposes, teachers may observe the following types of responses as children work through this unit.

at Level 3   Designing

- They generate ideas for greeting cards and realise that they cannot use the materials they had originally planned to use as they are too expensive.
- In discussion with the teacher they expand on their ideas.
- They explain how they would set about making the card.
- They make good use of the range of commercial cards available to generate ideas.
- They label their drawings to provide more information on the card they will make.
- They recognise that their design has to match the designs of the Diwali cards displayed.

Making

- They think ahead when making. They prepare the poem on a word processor while the stars they have stuck to the front of their card dry.
- They use scissors and trimmers accurately.
- They design and make a simple border to surround their poem as they feel that it will be attractive when the card is opened.

at Level 4   Designing

- They have visited a card shop at the weekend and talk about some of the ideas they saw. They use the ideas in their designs.
- They explain that one of their designs is made up of very strong colours and would not be suitable for the elderly person for whom it is designed.
- They evaluate their work and decide to change the card as the moving part they have attached looks untidy. They redesign the moving part to make it smaller.
- They use paper to model the pop-up moving part and decide that it would take up space that was to have been used for the message. As a result, they decide to use an alternative pop-up design they have also modelled in paper.

Making

- They make a sequence of drawings explaining the stages they will work through to make their card.
- They list the tools and materials to be used.
- They carefully and precisely cut out the various card parts to be attached to their card. They attach the pieces by sticking and by using paper fasteners to create movement.
- They decide to use different glues for sticking paper and felt.
- They explain that the overall image of their card is good and would appeal to an elderly person. However, they feel that the type they have used for the poem is inappropriate as an elderly person would find it difficult to read.

at Level 5   Designing

- They draw upon their knowledge of a range of different greeting cards and from their visit to a card shop.
- They discuss with the proprietor of the card shop what would be a suitable card for a teenager's birthday. They use the information gathered to inform their designs.
- They design their card so that, when a piece of card is pushed, a bird moves across into a birdcage. They modelled the part from paper prior to making it to ensure that it would work and that they were working to the correct scale.
- They decide that, although their design for a pop-up card is a good one, it is impractical because it would be too bulky to fit into an envelope.

Making

- They work entirely from their plans. They discover that the swinging monkey on the front of their card does not swing freely using a paper fastener. They go back to their plans and sketch out the changes they will have to make in order to create the swinging motion as the person receiving the card picks it up.
- They use trimmers and cutters to create the card parts they require.
- They use a computer to produce a rhyme to place inside the card. They work independently at the computer, being able to load the word processor, choose an appropriate typeface, place a drawing within the text and print out to the correct shape to fit on to the card.
- They use the tab facilities on the word processor to set the correct size and space in which to write their rhyme. When they discover that their rhyme is too long to fit on to the card, they change the size of the type.
- They evaluate their card against their design and explain that they could have made the parts of the moving part thicker by reinforcing each piece because it has snapped on several occasions when used.

BLACKPOOL ILLUMINATIONS

Context Children may gather information to inform their designing and making by visiting an illumination display, town centre lights, Christmas or Diwali light shows or an illuminated shop window display.

Outcome Making an illuminated display. Individual children or groups of children will make an illuminated display which can then be placed together to create a light show – their own 'Blackpool Illuminations'.

Cross-curricular links
SCIENCE Physical Processes – Electricity.

RESOURCES FOR INVESTIGATING
- A collection of photographs of any illuminated display. These may well be obtainable from local travel agents.
- A wide range of electrical items for creating circuits and making switches. Include:
  – 2.5V or 3.5V screw-in bulbs
  – 1.5V batteries
  – battery holders
  – black film canisters for use as battery holders
  – lengths of wire
  – 3V buzzers
  – 1.5V DC motors
  – wire strippers.
- A speed control switch kit. See pages 109–112 for details of electrical equipment.
- A visit to an illuminated display.

RESOURCES FOR MAKING
- wooden strips
- pre-cut card triangles
- a wide range of card – both thin and thick
- coloured plastic sheets
- a range of electrical items (see above)
- tape recorders
- sets of Lego Technic kits, including motor control mechanism

ACTIVITIES   Focused practical task

- Gather together a wide range of electrical items. Explain to the children how to create a simple circuit. Encourage them to link a number of bulbs into their circuits to observe the effect.
- Demonstrate how to create a battery holder from a film canister in order to ensure a good contact is maintained (see page 109).
- Construct series and parallel circuits that include three bulbs within each circuit. Ask the children to create another circuit that lights one bulb. Ensure that the bulbs they use are all 1.25V. Ask the children to observe

the difference in the brightness of the bulbs within the three circuits. They will notice that in a series circuit the three bulbs will each be a third the brightness of a single bulb in a circuit. In the parallel circuit they will have noted that each bulb is as bright as a single bulb would be. This is because each bulb is connected to the battery along its own circuit.

■ Introduce the children to the following ways of making bulbs brighter and dimmer.
  1. By connecting bulbs into a series or parallel circuit.
  2. By connecting a speed control switch kit into the circuit.

■ Teach the children how to cut and stick together wooden strips using card triangles (see pages 104–105). Make simple structures using this technique and test the strength of the different shapes. The technique is a good one to use in science to find out at what point a structure will break. Try leaving the card triangles out and simply gluing the wood. Test the strength of two structures of the same shape but with no card triangles in one.

Note: It is also possible to vary the light emitted from a bulb by connecting more batteries to the circuit or placing a different sized bulb within the circuit. However, this can prove expensive as the bulbs will blow if too many batteries are connected. For instance, if you use two 1.5V batteries, they will blow a 1.25V bulb, so it is better to use a 2.5V bulb if connecting more than one battery. It is a good idea to demonstrate how a bulb can blow to discourage children from attempting to use two 1.5V batteries without changing to a 2.5V bulb first.

Investigative, disassembly and evaluative activities

■ Gather together a collection of photographs that you have taken on a visit to an illuminated display and pictures of illuminated displays, taken from holiday brochures or magazines.

■ Observe the different effects created by the displays. Which do the children feel create the best effects? How do they think those effects were created?

■ Which of the lighting effects do the children think they could create? How would they create each effect?

■ Encourage the children to observe the colours used in the displays. Which colours are used most often? Are there some colours which the children feel create a special effect when used together?

■ Is sound used with the displays? What effect does it create? How do the children think they might include sound in their displays?

■ Identify types of movement that the children could incorporate in their illuminations and provide Lego Technic kits for children to explore how they might create such movement.

Design and make assignment

■ Decide on a theme for the illuminated display. The theme will provide the children with ideas.

■ Create an area where the illuminated displays are to be placed. This will give an indication of the scale to which the children are expected to work. Explain that they are going to place all their displays together in order to create an illuminated display in the classroom.

■ With the class, brainstorm ideas for displays and make a list that the children can refer to when designing.

■ Remind the children of what they have learnt about electricity and the lighting effects that they can create.

■ Ask the children to design their illuminated displays but explain that their drawings should clearly show the resources that they are going to need.

■ Where moving parts are to be incorporated, direct the children towards modelling the movement using Lego Technic. They should then incorporate the mechanism into their designs, showing clearly how it will be attached.

■ Making an illumination is a useful way of introducing children to movement and mechanisms, as any mechanism they create from Lego can be stood or attached to the rear of their hoarding and will be hidden from view.

■ Once the children's designs are complete, ask them to explain the order in which they are going to set about each stage of the task. Can they explain the circuit they are going to make and will it create the effect they expect?

■ Once the children are clear about what they are going to make and the order in which they are going to make it, they can begin making.

■ Emphasise throughout that the displays should be of the highest quality. Explain, for example, that they must ensure that the circuits are hidden away and not on view.

■ Throughout, gather the children together to evaluate each others' displays. This helps to maintain a standard of design and of the final product.

■ Encourage children to take note of the evaluations and make necessary alterations to their displays.

■ Some groups might like to add sound by using a tape recorder. They could consider which sounds or songs would best match their display. Some children may want to record their own music.

■ Once the children have made their displays they can be set up in a dark room. The children will enjoy evaluating the effects created by the individual displays as well as the effect created by combining them.

Extension activity

■ Children could be given the opportunity to link their illuminated display to a computer control system. They can link a number of bulbs to the interface box and then write a simple procedure to create a complex lighting effect. Children enjoy writing procedures to create a 'Mexican wave' effect or to get the lights to turn on in turn, backwards and then forwards.

ASSESSMENT

For assessment purposes, teachers might observe the following types of responses as children work through this unit.

at Level 3   Designing

• They make designs for their illumination and explain that they have to work with wooden strips to create a framework on which to place their illumination. They explain that their original design could not be made because they had not considered the frame to which it would be attached.

• They explain how they will set about making their illumination. They also explain their other ideas for illuminations.

• They plan to cover their illumination with Shireseal because they explain that some of the illuminations they saw were dirty and they want to ensure that theirs can be easily cleaned.

• Their sketches are clearly labelled which helps them to explain their intentions.

Making

• They make the framework first so that they have an indication of the scale at which they are working.

• They mark and cut the correct lengths of wooden strips and join them using card triangles to make a framework.

• They cover their illumination in Shireseal so that the light reflects off it and creates a better effect at night.

• The finished product is similar to their original design but they explain that they had to change the design because they could not attach a motor on to the thick card from which they had made the illumination.

at Level 4 **Designing**

- To assist their designs, they gather information from a range of travel brochures.
- They generate a number of ideas and can explain why they chose one idea in preference to another.
- The theme for their design reflects that they are aware that it needs to appeal to a specific audience. When questioned, they explain that the people who go to illuminations are children, families and the elderly.
- They evaluate their work and decide that it is not interesting enough. They explain that a good illumination has moving parts, sound and plenty of flashing lights. As a result, they rewire the lights so that they can create a flashing sequence, worked from a simple switch they have made.

Making

- They make sketches showing what they will have to do at each stage of making. Tools and equipment are listed at each stage.
- They explain that the framework and light system have to be made first and then brought together to create the illumination.
- They cut and join thin and thick card and use corriflute sheets for the front of their illumination.
- They reject some parts that have been made because they are not of the right quality.
- They explain that they have done a good job in hiding away all the wires but could have secured the bulbs better so that more light shone through when they were switched on.

at Level 5 **Designing**

- They write to the Blackpool tourist office, asking for information regarding the illuminations and use the information to inform their planning.
- They visit a set of illuminations and use what they saw to inform their designs.
- They spend a considerable amount of time discussing their plans to ensure that they are all clear about what they are making and in which order they will make each part.
- They use electrical components to create the flashing sequence they require, before attaching anything to their framework. They plan an alternative light sequence to use if the one they are modelling does not work.
- They devise a way of hiding the wires and the power source. They explain that this is because the illuminations must be safe if people are to be allowed to walk around them.

Making

- They work from their original plans but modify them when their light sequence does not work. They decide that the 'Mexican wave' they want to create with the lights would be easier to set up using a computer control program. They change their designs accordingly.
- They set up and organise the computer control system to write a simple procedure for creating the 'Mexican wave' effect.
- They model their mechanism for a 'swinging face' from Lego Technic and mount it at the rear of their hoarding.
- They evaluate their product and explain that they could have improved it if they had attached the lights last, after all the other moving parts were installed.

TRANSPORT

Context Children study a variety of transporters and link their design to what it is that they are required to transport. They will need to visit and see firsthand various vehicles and to observe the various parts of the transporters and the variety of designs. They will need to consider ways in which they can create similar shapes and moving parts from the equipment and materials available.

Outcome Making a transporter to transport specific items. The transporter will be made from boxes with wooden wheels and dowel for axles.

RESOURCES FOR INVESTIGATING
■ A range of model transport vehicles that children can bring to school.
■ A visit to look at transport vehicles.
■ A collection of photographs of different transport vehicles.

RESOURCES FOR MAKING
■ A collection of boxes of different sizes. Shoe boxes are an ideal size. If boxes are unsuitable for painting or gluing, the children can stick a layer of white paper over them or turn them inside out so that any printing is on the inside. Layers of white paper can be added to the model at the start or once the vehicle is constructed.
■ A wide range of electrical items for creating circuits and for adding lights to the vehicles. Include:
– 2.5V or 3.5V screw-in bulbs
– 1.5V batteries
– battery holders
– black film canisters for use as battery holders (see page 109)
– lengths of wire
– 3V buzzers
– 1.5V DC motors
– wire strippers.

ACTIVITIES

Focused practical task

■ Children should be taught and have the opportunity to explore:
– How to make a simple vehicle frame by punching holes into a box and then passing dowel through and adding pre-cut wooden wheels to the ends of the dowel (see Figure 3.1).
– How to create a simple circuit which includes more than one bulb.
– How to make a simple switch to incorporate into their circuit.
Figure 3.2 shows how children can make a chassis from wooden strips, dowel, switches and a simple circuit.

Investigative, disassembly and evaluative activities

■ Provide opportunities for children to observe a variety of different transport vehicles and to consider their design in relation to what they transport.

Figure 3.1

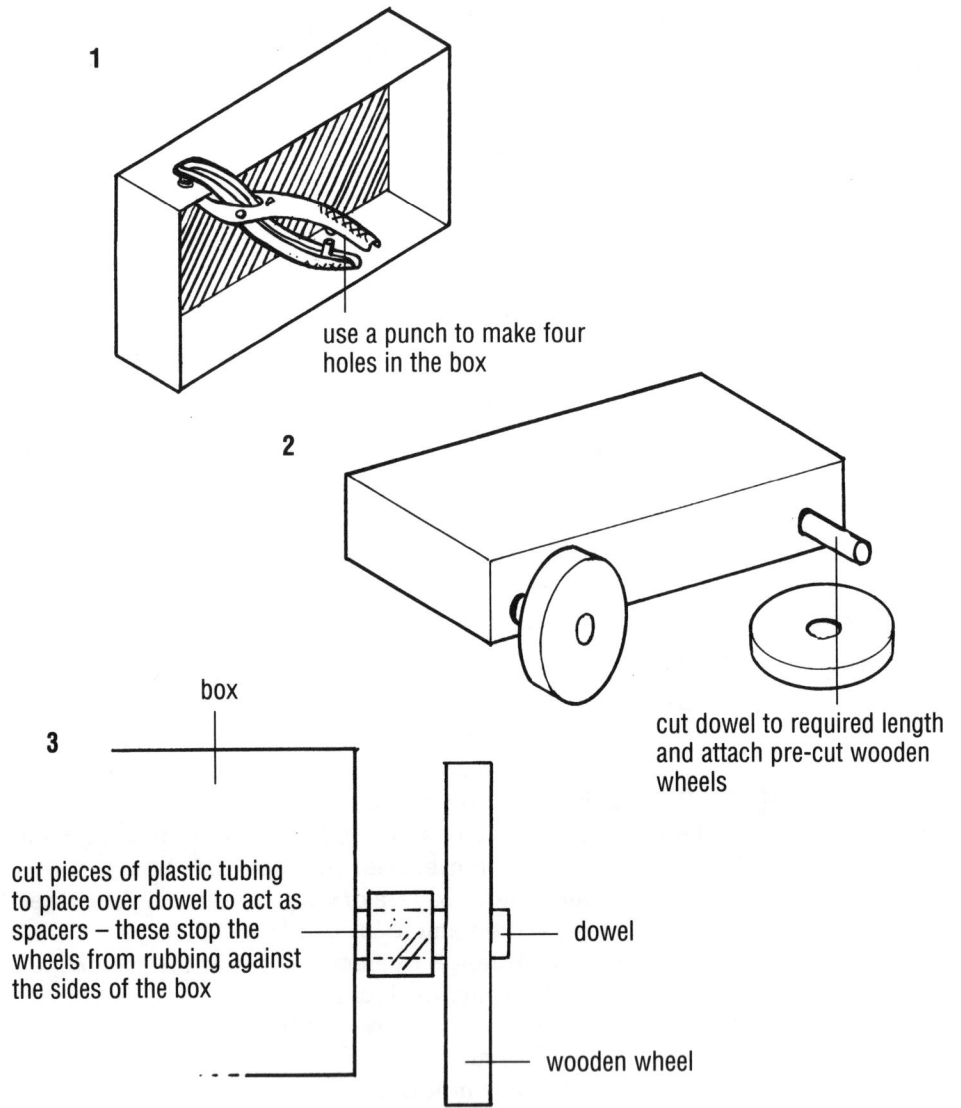

1 use a punch to make four holes in the box

2 cut dowel to required length and attach pre-cut wooden wheels

box

3

cut pieces of plastic tubing to place over dowel to act as spacers – these stop the wheels from rubbing against the sides of the box

dowel

wooden wheel

Figure 3.2

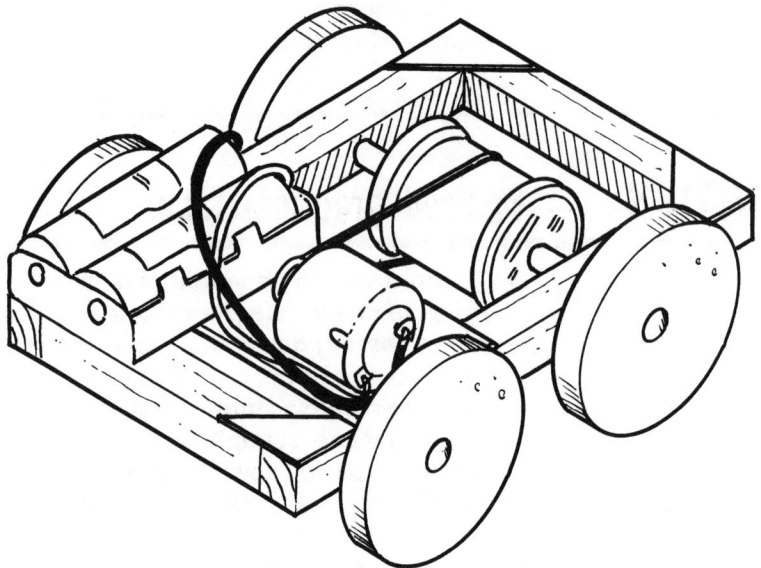

- Make a list of a variety of products that are regularly transported and ask the children to consider the design of the vehicles that would transport each one. The following list might prove a useful starter:
  - animals
  - grain
  - liquids
  - boxes
  - sand/gravel/building materials
  - food.
- Explain that many products are boxed so that they can be packed into a transporter more easily.

Design and make assignment

- Work with the whole class on considering the different types of vehicles that could be designed and the products that they would transport. Make a list of the children's ideas for them to refer to when designing.
- Remind the children of the method of designing a vehicle from boxes. This is important as it will ultimately affect their designs.
- The children should draw details of their designs. Some may draw the vehicle and label clearly to show the materials to be used; others might do a series of drawings showing each stage of the making of their vehicles.
- Once the children finish their designs, ask them to talk through their ideas, explaining clearly the order in which they will set about tackling each stage of making. They should be clear at which stage they are going to add the lights.
- Once the children are clear about how they are going to set about making their vehicles, they can begin making.
- Ensure that the children take responsibility for their working area. Explain the health and safety aspects involved with the equipment and resources they use in their making.
- Encourage the children to evaluate their vehicles at each stage. Their evaluations should lead to modifications to the design of their vehicles.

EXTENSION ACTIVITY

- The children can design a logo for their transport company and consider its size and location on their vehicle.

ASSESSMENT   For assessment purposes, teachers may observe the following types of responses as children work through this unit.

at Level 3   **Designing**
- They generate a design for their vehicles and explain that, although they originally wanted the lights up on the cab roof, they were not placed there because they would reflect into the eyes of other drivers.
- They explain how they will set about making their vehicle and explain other ideas that they have for a vehicle.
- They consider their vehicle's appearance by designing a colour scheme and a logo.
- Their designs are clearly labelled.

Making
- They explain the order in which they are to make each part. They consider the sequence for making.
- They make simple reflectors for their lights so that the light is directed forwards.
- They cut and join wood strips to make a chassis for their vehicles. They drill holes in the correct places for the dowel axles.
- Their finished vehicle looks similar to their original design but they explain that they changed the design for the carrier section because they realised that a round tube shape was difficult to attach to the wooden chassis.

at Level 4   Designing

- They gather together photographs from magazines as a reference for generating ideas.
- They recognise that their vehicle for transporting sheep around a farm will need to have soft material placed around the sides to protect the sheep in transit.
- They evaluate their vehicle as they are designing, taking into account what the vehicle is to be used for. The dumper truck will need to be able to dump its load as close to the site as possible, so the children provide it with taller wheels so that its chassis is clear of any obstacles at ground level.
- They make several sketches of the positioning of an oblong container lorry. They decide that if it is to be set on its side then it will be less streamlined and use more fuel. They recognise that there are few things that they can think of that need to be transported upright in a container.

Making

- They make sketches showing what they intend to do at each stage of making.
- They list the tools and materials to be used.
- They explain that they will either need to use plastic sheeting or paint a large sheet of card with Marvin in order to make it waterproof, which will allow them to use it as a container for transporting water into villages in times of drought. They model the waterproof tank prior to making it to the correct size to ensure that the Marvin method works and that the correct shape can be constructed from the card and that it will remain waterproof.
- They take great care to ensure that their finished vehicle looks right and functions correctly.

at Level 5   Designing

- They visit a transport company and look closely at the different vehicles. They collect photographs of various vehicles and write to several companies for information. They use this information to inform their designs.
- They discuss their ideas until they are satisfied that they are all clear about what they are going to make and the order in which they will make each part.
- They decide to place a buzzer on their vehicle to warn people when it is dumping materials.
- They can be heard discussing each idea for a vehicle in terms of its function and where it would be most likely to be used. The ideas discussed inform their designs.

Making

- They model the lights and battery holder for the vehicle before finalising plans, so that they can gauge where to site the electrical components. They change their design to ensure that a casing for the battery holder is incorporated.
- They use all the tools correctly and with confidence and take care to ensure that measurements are accurate.
- They evaluate their vehicle and explain how it matches their original design. They explain that they should have waterproofed again in the folds of the card they had covered in Marvin. It currently leaks slightly and the card has gone too soggy to repair.

ACTIVITY CENTRES

Context The evaluation of a range of activity centres made for young children. Children will evaluate the different moving parts and the effects created. They will link the ideas they observe to the ways in which they might be created with the range of materials that they are familiar with. They will evaluate how far the activity centres stimulate young children's senses.

Outcome Making an activity centre for a young child that stimulates the senses.

Cross-curricular links
SCIENCE The making of activity centres could be linked with work on forces in science.

RESOURCES FOR INVESTIGATING
- A range of activity centres for the children to evaluate, to promote ideas that can be incorporated into their own centres.
- Children can bring in their old centres or their younger sisters' or brothers'.

RESOURCES FOR MAKING
- A range of medium sized boxes that can incorporate all the moving parts for the centres.
- Card of different thicknesses.
- Paper clips and fasteners.
- String and cord.
- A wide range of electrical items for creating circuits. Include:
  - 2.5V and 3.5V screw-in bulbs
  - 1.5V batteries
  - battery holders
  - black film canisters for use as battery holders
  - lengths of wire
  - buzzers
  - 1.5V DC motors
  - wire strippers.
- A range of pneumatic equipment. Include:
  - a variety of different sized syringes
  - plastic tubing to fit on the end of the syringes
  - two- and three-way connectors. See pages 112 and 114 for further guidance on pneumatic and hydraulic equipment.
- A set of Lego Technic kits.

Health warning
- Only new sealed syringes purchased especially for work on pneumatic and hydraulics should be used. Explain to the children that they should never pick up syringes that they see out of school.

ACTIVITIES Focused practical task

- Ensure that all the children can make a circuit using batteries, wire and bulbs. They should be shown how to use a black film canister to create a battery holder (see page 109).

- Introduce buzzers and motors and allow time for the children to learn how to make firm connections. More able children may experiment with ways of slowing down the motor and others may use Lego motors and gears to create different forms of movement.
- Make coloured wheels and attach to the ends of a 1.5V DC motor using Blu tac. Several activity centres use such an effect to create spinning colours.
- Demonstrate how tubing attached to different sized syringes can create movement. Use the connectors to connect more syringes and allow time for children to explore the effects created (see page 114).

Investigative, disassembly and evaluative activities

- Make a collection of activity centres and observe the different effects created. Explain that the effects created appeal to the young child's senses.
- Working in groups, the children could take an activity centre and make sketches explaining how they think the effects are produced. Their sketches should contain enough information to make it clear to others how the effect is created.
- Encourage the children to relate the effects they observe to the focused practical tasks they undertook.
- Ask the children to identify the effects that they feel they could incorporate in their own activity centres.

Design and make assignment

- Ask the children to note down ideas that they are going to use in their activity centres. On a small sheet of paper, they can begin to decide upon the positioning of each effect they are going to make. Which senses do their effects appeal to? Have they decided upon effects that will appeal to a sufficiently wide range of senses?
- Children should now begin to make more detailed drawings of their activity centres. They may have to 'explode' some parts to explain in more detail how they are going to create and attach the effect.
- Children should be provided with the opportunity to talk through their ideas with the teacher before they begin making. It is important to check that each group has planned the order in which they are going to make each part and how they are going to attach it to their centre.
- The children should be encouraged to model any moving parts using paper, card or Lego Technic, as appropriate. Mechanisms that they make from Lego Technic may be incorporated into their activity centres.
- Once the children are clear about how they are going to set about making their centres, remind them that they should consider the quality of the finished article, as nobody is going to want to use a centre that is not attractive to look at. Also remind them of the safety considerations if their centres are designed for people to use. At this point, some groups may wish to make further changes to their designs.
- The children should be encouraged to evaluate their centres at each stage of making and to make changes where they feel it is appropriate.
- At appropriate stages bring the whole class together and encourage them to positively evaluate the work of others and to suggest ways in which the work may be improved. Point out any very good or original ideas and encourage others to use them if appropriate.
- When the activity centres are complete, ask the children to say how they might evaluate their work. They may suggest that the centres should look attractive or that they are suitable for a certain age group or that they appeal to a wide range of senses. Make a list of their ideas and encourage each group to evaluate the work of others, using the criteria that they have set themselves.

■ If the activity centres have all been designed for young children, the reception class could be invited to carry out an evaluation of the work. Afterwards, ask the children if the reactions of young children were surprising. Ask them whether they can explain why this might be.

EXTENSION ACTIVITIES

■ Children may wish to design an advertisement for their activity centre. Gather some toy adverts and analyse the information that is given. Encourage the children to use the information to design and make their own adverts.
■ Children can make attractive boxes for their centres.

ASSESSMENT　For assessment purposes, teachers might observe the following types of reactions as children work through this unit.

at Level 3　### Designing
• They generate a design for an activity centre.
• They realise that, although many of their ideas would be attractive to a young child, they would not be safe.
• They explain how they will set about making their activity centre and display a wider range of ideas than those they use in their final design.
• They look at the activity centres on display in the classroom and use some of the ideas in their own designs.
• They label their designs to aid clarity.

Making
• When making, they think ahead and suggest that they will need to attach each item to their box and then paint the whole thing, because attaching items afterwards would ruin their painting.
• They choose the most appropriate tools and materials for the task.
• When everything is attached, they cover it all with white strips of paper and paste, so that it will be easier to paint.
• The finished product looks similar to their design.

at Level 4　### Designing
• They gather activity centres from home and cut out photographs from toy magazines. They use the information to inform their designs.
• They carry out a survey by taking the activity centres they have gathered to a nursery and asking a number of parents which parts of each activity centre they like best.
• They evaluate their work as they proceed, deciding that they will have to enclose the small ball in a tube, as a small child could swallow it. They change their original design accordingly.
• Their designs illustrate a wider set of designs and ideas than are finally used.

Making
• Step-by-step plans are produced, clearly showing each stage of making.
• All tools and materials to be used are listed.
• They use a variety of cutting and shaping tools to create the shape and size they need from a range of everyday items made from card and plastic.
• They use a variety of different glues, tapes and fasteners to attach the various pieces to their boxes.
• The quality of the finished activity centres indicates that they have been made for the purpose of being used by others.
• They identify how well the ball shoots up the tube when you blow through a tube, but explain that they need to think of another method for blowing air into the tube, as using one tube is unhygienic.

at Level 5 **Designing**

- They make a collection of photographs of activity centres from various magazines. They visit shops with their parents and gather information regarding activity centres for children and adults.
- They talk with the reception children, asking them what they would like to see on an activity centre designed for them.
- Their research has informed their designs.
- Their designs show an understanding of the use of syringes and tubing to create controlled hydraulic movement.
- They model the small ball shooting through a tube using a clear plastic tube, a balloon and a clear plastic container. As a result, they adapt its size so that it will fit on to their design for an activity centre.

Making

- They make their activity centre from their original design. They have adapted their design, having modelled one of the moving parts and discovered that it needed to be larger to create the right effect.
- They model a winding mechanism from Lego Technic. They discover that it wound up too quickly so they change the gearing to slow it down. They incorporate their working mechanism into their activity centre.
- They use appropriate tools and materials accurately and with confidence.
- They invite the reception children to evaluate their activity centres. They observe which parts of the centres they use most and then explain to their own class the results of the evaluation.

YOGHURTS

Context

To undertake an analysis of a variety of yoghurts in order to establish the children's taste preferences. To use the information gained to assist them in designing and making a yoghurt.

Outcome

Making a yoghurt.

Cross-curricular links
HEALTH EDUCATION
& SCIENCE

This could form a link with work being undertaken on the senses or health.

The children could undertake work on different types of foods in order to develop an understanding of which foods are considered to be healthy for us and why. Such an understanding will inform their thinking when making choices about the ingredients they will use.

Health and Safety

Before undertaking any work with food ensure that:
- You are fully aware of any dietary requirements that may influence the foods that you can make available to the children.
- You are fully aware of any health and safety requirements set down within the school's Design and Technology and Health and Safety Policies.

When undertaking work with food:
- Ensure that a specific food area is established and that all tools and clothing used are for use within food activities only.
- Ensure that all children are fully aware of any rules regarding personal hygiene.

For further guidance:
Refer to the sample Design and Technology Policy, pages 26–30, which provides specific guidance on health and safety matters relating to the teaching of food technology.

RESOURCES FOR MAKING

- natural yoghurt
- dried apricots and apples
- food colourings
- refined sugar

This is a basic list of ingredients that can be added to to match the activity to the age and ability of the children. The degree to which the ingredients list is expanded is likely to be influenced by the degree to which the children have previously undertaken work with food.

Other ingredients may be added to the list that form a link with work the children may be undertaking on fruits or in a healthy eating project.

Fresh fruit can be included which will involve the children in setting up an area complete with cutting board and tools in order to prepare the fruit for combining with the yoghurt.

ACTIVITIES   Focused practical task

- The children should be shown how to set out an area in which they are going to work with food. The need for personal hygiene should be emphasised with particular reference to tasting. It is a good idea to provide two different coloured spoons. The children should be aware that one colour denotes spoons used for collecting food and placing it on to paper plates or into their yoghurts and the other colour denotes spoons for tasting foods. They should ensure that they always use their own plastic spoon for tasting, which will be collected and thrown away at the end of the activity.
- Decide which aspects of a yoghurt you are going to ask the children to focus upon – its texture, how healthy it is, its colour, how sweet it is, whether it is crunchy. If children are to consider designing a healthy yoghurt then they will need to be able to classify the ingredients that are made available. It is important that the children are familiar with the ingredients which you make available to them.

Investigative, disassembly and evaluative activities

- The following are useful resources for setting out a **tasting** area:
  - white plastic spoons for tasting food (to be disposed of afterwards)
  - plastic bag for placing used spoons in for disposal
  - coloured plastic spoons for scooping food onto paper plates
  - plastic bag for collecting together coloured spoons for washing
  - paper plates and plastic cups
  - kitchen towel
  - plastic sheet to cover tables
  - straws.
- Gather together a selection of different yoghurts for the children to taste. Children can use the name of the yoghurt that they find on the label or each can have a letter stuck on to the side for ease of reference. Set out the tasting area so that children can scoop out small amounts of yoghurt to place on to their paper plates for tasting.
- Provide each child with a Tasting Food Sheet (Appendix 4) so that they can record their likes and dislikes.
- After tasting each yoghurt each child should complete their Tasting Food Sheet and then take a drink of water to clear the mouth for the next tasting.
- Make a large blank graph for the children to enter which yoghurt they liked most.
- Discuss with the children their likes and dislikes. Encourage them to explain why they liked a particular yoghurt in terms of its colour, texture, flavour, name etc.
- Make a collection of the labels from several yoghurts and place them on display. Identify the type of information that is set down on every label. Discuss with the children which labels they like best. Extend the work by linking labels to specific groups of people. Which label would appeal to a very young child, a teenager, an elderly person or somebody on a diet?

Design and make assignment

- Recap on all the work covered and remind the children of their preferences. Decide who the yoghurt is to be designed for.
  - themselves
  - a young person
  - an elderly person
  - a teenager
  - for a party
  - to link with a theme, e.g. a film or book.

- Once you have decided who the yoghurt is to be designed for, then discuss with the children what they think may be the preferences of that particular person or group with regard to the contents of the yoghurt.
- Provide the children with a Star Diagram (Appendix 5) and explain that it will help them design their yoghurt. Ask them to list at the ends of the star what they think designers would have to consider when designing a yoghurt. They should consider all factors that they feel are important and may include: that it tastes good, is crunchy, is healthy, is fruity, has an attractive label, has an original name, is creamy or is brightly coloured.

Their Star Diagram may look like this:

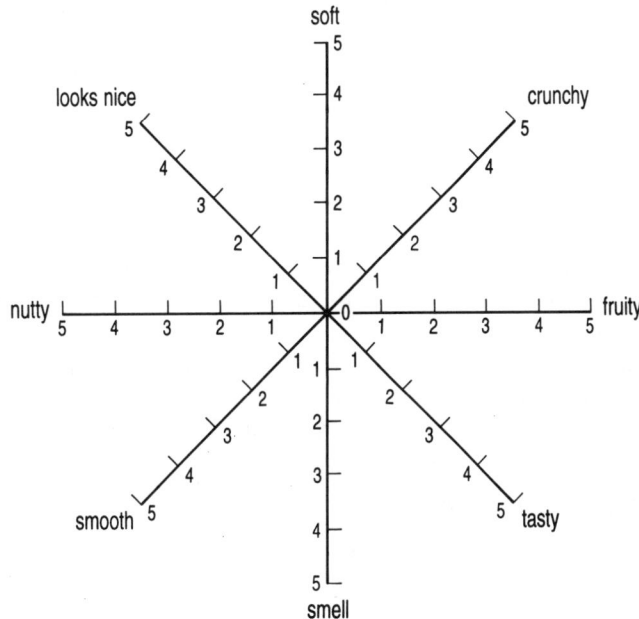

- Now ask the children to grade how important they consider each of their listed qualities is on a scale from 0 – not at all, to 5 – very. Once they have completed this task then ask them to join up their points to make up their star.

Their Star Diagram will look something like this:

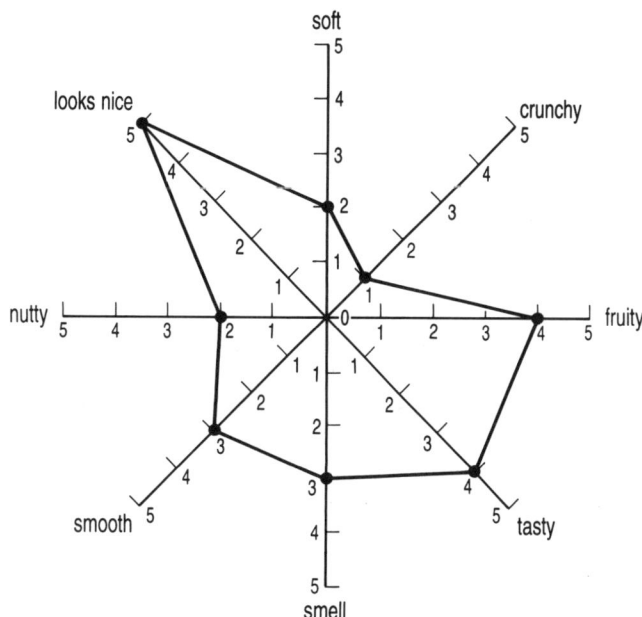

- Explain to the children that the star indicates what their yoghurt should be like once they have made it.
- Set out all the ingredients that the children are to have available. Bring each to the notice of the children.
- Remind the children of any health and safety factors concerning personal hygiene. Explain how they are to set out their working area so that they work in an organised way with due attention given to health and safety factors.
- The children can now set about making their yoghurts. Once the children have made their yoghurts others should be provided with opportunities to taste each in turn.
- Prior to tasting provide each group with a new blank Star Diagram and ask them to note down the qualities that they had originally set down around the previous Star Diagram. As each child tastes the yoghurt they should place a tick around the star to show how much they consider the yoghurt meets each criteria listed.
- Once everybody has completed the Star Diagram then each group can indicate the most popular response from 0 to 5 along each line. They can then join the points and create the Star Diagram which indicates the views of the whole class. They should then consider how far the star fits their original one. In other words if the two stars are similar in shape then the class agree that you have produced a yoghurt that met your original specification.
- Where the two stars are very different the group should make a note and explain what they might do, if they were to make another yoghurt, that would ensure that those particular criteria were met.

EXTENSION ACTIVITY

- The children can use the information they have gained from observing the collection of yoghurt labels to design their own. They can design and make a yoghurt label to fit around a plastic cup. They can then be displayed with information explaining who the yoghurt was designed for. Other children can evaluate how they consider that the label would be suitable for the group of people it was designed for.

ASSESSMENT

For assessment purposes teachers might observe the following types of responses as children work through this unit.

at Level 3 **Designing**
- They generate a design for a yoghurt and recognise that they cannot have a creamy and a crunchy yoghurt.
- When questioned, they suggest other ingredients that they might use to make their yoghurts creamy.
- They plan to make their yoghurt a deep red colour as they think this will make it more appealing to young children.
- They create their Star Diagram.
- They can explain the order in which they are going to set about making their yoghurts.

Making
- They colour their yoghurt before placing their dried fruit into it.
- They use the tasting equipment correctly.
- They grate the nuts so that the yoghurt is creamier than if they had placed the nuts in whole.
- Their yoghurt does not match their Star Diagram but they explain that they had added more sugar because they thought it originally tasted too sour.

at Level 4 **Designing**

- They have a number of ideas based upon information they have collected from tasting the yoghurts and from the group's Star Diagrams.
- They explain that their yoghurts could be made healthier but they have decided to make their yoghurt from ingredients they like best.
- As they are mixing their ingredients together they decide that it will create too rough a texture so they decide to use fewer nuts and to grate them.
- They have several ideas for healthy yoghurts but explain that their choice falls between healthy yoghurts and their own preferences that do not match. They discuss each other's Star Diagrams and note the differences. They draw a new Star Diagram that they feel takes into account all their preferences.

Making

- They explain each stage that they will work through to make their yoghurt. They also explain the order in which they will undertake each step.
- They weigh each set of ingredients so that, if they should need to make another one, they can see the amounts they have used.
- They measure and weigh out the ingredients accurately having chosen appropriate scales for the task.
- They set out their yoghurt for tasting on a place mat so that it looks more appetising. They take care at each stage to ensure their yoghurt is presented well.
- They evaluate their yoghurt and explain that although they have created the colour of yoghurt that they planned the children in the class do not like it. They explain that they will need to create a less bright colour the next time.

at Level 5 **Designing**

- Their designs are clearly drawing upon ideas generated by their evaluation of the commercial yoghurts they tasted. The yoghurt design is closely matched to their likes of commercial products. They choose a light colour for their yoghurt because most of the commercial ones they observed were light in colour.
- Their designs are based upon all the information they have collected together from commercial products, their Star Diagrams and the stars created by others in their group.
- To create the creamy taste of the commercial yoghurt that proved most popular they choose their ingredients with care.
- They mix small amounts to begin with so that they can taste and evaluate before deciding to make larger amounts for their yoghurt.

Making

- They work with constant reference to their Star Diagrams throughout the making stage to clarify what they are doing. They choose ingredients that they can cut and mix with the yoghurt to ensure it maintains its creamy consistency.
- They choose the appropriate tools and weighing scales throughout the making stage.
- They are not happy with their first mix because it tastes too strongly of orange. They keep a record of each ingredient so they can reduce the amount of orange juice they use next time. They make the changes and produce a yoghurt more to their liking.
- As their yoghurt was designed for adults they invite parents to taste their yoghurt and to complete a Star Diagram. They make comparisons between Star Diagrams and explain where changes might be made to produce a better match to the adults' preferences.

PACKAGING

Context Children will study and evaluate a variety of packages, looking carefully at all aspects including nets, designs, protection and printed information. They will evaluate the suitability of the packages to the contents they are designed to hold.

Outcome Making a package for a specific product.

Cross-curricular links
MATHEMATICS

Shape, space and measures:
1a. Use geometric properties and relationships in the solution of a problem.
1e. Apply their measuring skills in a range of purposeful contexts.
2b. Make 2D and 3D shapes with increasing accuracy, recognise their geometrical features and properties, and use these to classify shapes and solve problems.

RESOURCES FOR INVESTIGATING

- A collection of packages. This can be a general collection of interesting packages or a number of packages based around certain products e.g. types of food, washing powders, toys.

RESOURCES FOR MAKING

- A range of different coloured card of varying thicknesses.
- Sheets of plain white paper.
- Various glues suitable for sticking paper and card.
- A strimmer for use by children.
- A range of resources for adding designs and colour to the packages created.

ACTIVITIES   Focused practical task

- Show the children how to measure, score and cut from card a net, complete with flaps, to create a 3D shape. Explain the value of drawing the net on the card in a position where it will create the least amount of waste.
- Provide an opportunity for children to create a design on a net prior to making it into a 3D shape. This should be undertaken on paper so that children become aware of the need to model a net prior to making it from more expensive squared card.
- Demonstrate ways of creating drawings and text on the computer. Show them how the size can be altered or the shape changed to make an exact fit for their package. Allow the children the opportunity to create text at the computer and to cut and mount it on their paper net prior to assembling into a 3D shape.

Investigative, disassembly and evaluative activities

- Explore the different shapes of the packages in relation to their contents. The children will soon realise the need for the packaging to tessellate in order to make full use of the space available when placed in large boxes for transportation.
- Allow the children to choose one of the shapes and to sketch out what they think the net will look like. More able children will be able to create the net to the exact size. The nets can be drawn on paper with tabs added and then cut out and glued together so that children can check whether they have created an exact replica of the package shape. The children can then open out the original package and compare it to the one they have drawn or made.
- Try to obtain a number of the same boxes to show how they fit together to make the best use of the space available.
- Ask the children to look at the information on the packages and draw up a list of the types of information that must be given and other information that is provided. Encourage them to look at the colours and lettering on the boxes and evaluate the attractiveness of the box and the clarity of the lettering.
- The children will soon think of other aspects of the box that they consider to be important for sales, storage or transportation. They may consider that certain colours can be used with clothing products but not with foods, that certain products need to be seen inside the box and others do not.
- If possible, arrange a visit to a factory or packaging plant where the children can follow the process of packing items into boxes and transporting them in bulk.
- A visit could be made to a graphic design agency, to discover how the designs for packages are created.

Design and make assignment

- Decide upon what the children are going to design and make packaging for. The following are tried and tested ideas:
  - Packaging a number of products for a company that is selling four for the price of three.
  - Designing and making a package to show off new netball or football kits that have been presented to the school by a local sponsor.
  - Providing the packaging for a mug or cup and saucer.
- Children will need to explore whether they are to design packaging for the safe transport of an article or packaging for display purposes, or both. The activity can be made as easy or as simple as desired.
- Encourage the children to gather examples of the product from local shops or from home. They can evaluate the packaging in terms of visual appearance, information provided, safe transportation and display. Encourage them to identify ideas that they might incorporate into their own designs.
- Children need to decide whether their packaging needs to cushion a fall and ensure the goods are not damaged. If so, they should consider the different materials they could use inside the package to protect the contents. Encourage the children to consider ways in which materials can be shaped to create suitable protection, for example, paper can be cut into thin strips.
- Ask the children to set about designing their packaging. They could come up with several ideas and then carry out some market research to find out the views of others on their different designs. Emphasise the need for accuracy. Children will need to take careful measurements of the product if they are going to make a net the correct size to hold their product.
- The children can model their packaging out of paper to check that it is the correct size and shape. This will save on wastage. Make sure the children

keep their models so that they can explain to others how they set about creating their package.

■ Before the children begin making, ask them to include in their designs a note of the order in which they will tackle the making of their packaging. At the design stage children should have given thought to the colour of the packaging and any information that is going to be printed on it. Check that the children have considered at what stage they will add colour and any writing. It is often easier to do this when the net is flat than when it has been constructed.

■ Once the children are clear about their ideas and have set down the order in which they are to tackle each stage, they can begin making.

■ At appropriate times, question the children about what they are doing and encourage them to evaluate their work. Encourage them to refer to their designs and to evaluate their packaging against their original plans. Children should make changes to improve their packaging wherever they feel it is appropriate.

■ The whole class can be gathered together to listen to several groups explaining what they have done. The class can be given the opportunity to evaluate the packaging in a positive manner. Point out work of quality to encourage other groups to aspire to such standards. Encourage the children to use good ideas from their evaluation of the work of others.

■ When the work is finished, ask each group to evaluate their work. They should be reminded of what they set out to design and make and asked to evaluate whether they have achieved their original intentions. Ask the children to consider how well the packaging works, for example, would it allow for the safe transportation of the product?

EXTENSION ACTIVITIES

■ Link with a local company who can supply several products for which the children can design packaging. Afterwards, the children can compare their own designs with the packaging that the company uses.

■ The packaging activity can be linked to work on food. The children can design a package for a sweet or a healthy eating bar that they have made.

ASSESSMENT For assessment purposes, teachers might observe the following types of responses as children work through this unit.

at Level 3 **Designing**
• They explain how they will set about making their package. They talk to the teacher about ideas that they discounted for various reasons.
• They explain that their package will need to be taped or glued at the ends to ensure that the product does not fall out.
• They set down their plans in the form of a flow chart, showing what they are going to do at each stage. They make a labelled drawing of the box and the parts of the design.

Making
• They begin to choose the most appropriate tools and equipment for themselves.
• They score and cut with some accuracy and add a double thickness to part of the package in order to strengthen it.
• They draw, cut out and assemble a net to create their package.
• Their package looks similar to their design. They explain that they changed the lettering so that it would not overlap on to more than one side because, when the packages are laid out side-by-side, the name of the product would be obscured.

at Level 4 **Designing**
• They gather a number of different commercial packages for the product that they have chosen to package. They make several different designs, drawing ideas from the packages they have collected.

- They explain that customers would want to see that the garlic inside was fresh, so they have incorporated a clear front panel into their design.
- At regular stages, they place their product into their package to undertake an evaluation. As a result, they discover that they need to place stops at each end to make sure that their product does not slide down the package and out of view.
- They make several designs and explain that the one they wanted to make had too many pieces and, ideally, packaging should be made from one piece of card.

Making

- Their plans indicate the order in which they will tackle the making of the package. They clearly label each part of the final design and list all the equipment and tools required.
- They cut out their net and choose an appropriate glue from a range available.
- They create drawings and text on the computer and change the size to fit the package. They take great care to ensure that the package looks as good as those seen in the shops.
- They coat the inside of their package with Marvin to waterproof it, in case the tomatoes inside become damaged.
- When they evaluate their package, they explain that, although they waterproofed the inside, they should have placed a section of a paper towel inside to soak up the liquid from any damaged tomatoes.

at Level 5   Designing

- Their designs clearly draw upon their evaluation of commercial packages. They set out the text on the side of the package in the way described to them when they visited the graphic designer.
- They design a package but explain that the shape is too complex and will not fit together with others when packaged for transportation.
- They model their package using paper prior to making from card.
- They choose a less attractive design because they discovered that the more attractive design that they wanted to make wasted too much card when they drew it out and modelled it from paper.
- They make a drawing of several of their packages to show how they would fit together. From their drawing, they work out the sizes of boxes required to transport 100 of their packages.

Making

- They work from their plans, referring to them throughout to check which stage they are at and that their package is turning out as intended.
- They use scoring tools and the strimmer with accuracy throughout.
- They use the computer to create drawings and text and alter them to the exact size required.
- They check their measurements and the placement of the computer-generated text and drawings.

MAKING A MEAL OF IT

Context

Children will evaluate the basic items required for a variety of different types of meals. They will gain knowledge of the costs of various ingredients required for specific meals and use this information to design and make their own meal.

Outcome

Making a meal for a specific purpose from a specified budget.

Health and Safety

Before undertaking any work with food ensure that:
■ You are fully aware of any dietary requirements that may influence the foods that you can make available to the children.
■ You are fully aware of any health and safety requirements set down within the school's Design and Technology and Health and Safety Policies.

When undertaking work with food:
■ Ensure that a specific food area is established and that all tools and clothing used are for use within food activities only.
■ Ensure that all children are fully aware of any rules regarding personal hygiene.

For further guidance:
Refer to the sample Design and Technology Policy, pages 26–30, which provides specific guidance on health and safety matters relating to the teaching of food technology.

Cross-curricular links
HEALTH EDUCATION
& SCIENCE

This unit can be linked with the Our Body project or other health education topics. It is important that children undertake work on food with a basic understanding of healthy eating and which are healthy foods and why.

This unit of work can be adapted to suit the needs of individual classes and integrated into a term's project. The unit can be adapted to include designing and making meals such as:
– a picnic for a summer's day
– sandwiches for a vegetarian
– a light, healthy snack for an elderly person
– an open sandwich.

RESOURCES FOR INVESTIGATING

■ A collection of menus which divide the meal into three courses.

RESOURCES FOR MAKING

■ Ingredients. Decide beforehand what ingredients the children will be able to choose from to make their meal. It is not necessary to cook anything, as the list can be designed to include only ingredients that can be used uncooked, within a classroom environment. A suitable list might include:
Main course – fruit and vegetables, bread.
Sweet – a range of fresh fruit, instant whip type dessert mixes.
Drinks – cordials and fresh fruit drinks.

If the work is to be linked with a project on healthy eating, the ingredients might include a range of foods of varying health value.

Each article can be priced by the teacher to encourage the children to make decisions about what to make in order to stay within budget.

■ A range of suitable tools and equipment for preparing the foods the children will choose from. Include equipment that will allow children to change the shape of the foods, such as graters and safe slicers.
■ A range of plates, cups and eating utensils that are safe to use.

ACTIVITIES

Focused practical task

■ Provide opportunities for children to classify foods according to how they contribute to a healthy diet.
■ Ensure that, before starting to design their two course meal, the children are familiar with all the equipment available.
■ Show the children how to set up and prepare an area for working with food, emphasising the need to ensure that they have clean hands at all times.

Investigative, disassembly and evaluative activities

■ Provide the children with opportunities to cut, shape and arrange the various foods to be made available. Emphasise the safe and hygienic way of working and the importance of presentation and colour.
■ Make a collection of menus and relate the dishes to the children's knowledge of healthy foods.
■ Provide an opportunity for the children to taste and evaluate the various foods. They can classify them according to their texture, flavour and health value.
■ Use the Star Diagram (Appendix 5) so that the children can begin to focus on the main factors that they have to consider when designing and making their meal. Explain how to use the star and that it will provide information for them to refer to when designing their meal. Children may consider that the main criteria will be looks, healthy ingredients, taste and cost which they can place around their Star Diagram.
■ Children could use the Tasting Food Sheet (Appendix 4) to record their preferences for the different ingredients. When children are working in pairs or groups they can use the completed Sheets to help them decide which ingredients are most popular.
■ Prepare a survey of children's preferences using Junior PinPoint or Data Sweet. The children can graph their results.

Design and make assignment

■ Discuss with the children all the things they will need to consider when designing their meal. Make a list of their comments, which should clarify whether:
  – it is a two course meal
  – it is a healthy meal
  – the ingredients can be bought within the set budget (if you decide to set one)
  – it can be made in the time available
  – the children have to set out and prepare their working area
  – the meal should look attractive
  – they know all the safe and hygienic ways of working at all times.

- Show the children the food, utensils and equipment available.
- Ask the children to design their meal, including details of how they will make the meal, the order in which it is to be made and the ingredients they will use.
- Once the children have thought everything through allow them to begin making the meal.
- Once the meal is made, the children can show the rest of the class who can make positive comments on how it looks and what they would most like to eat.
- Ask the children to evaluate their meal before they set about tasting it. How might they have improved it?
- Once the children have tasted their meal, ask them once again to evaluate it.
- Carry out a survey of the key nutrients of the meal related to healthy eating. Children can draw upon the information provided on the food label or from data materials.

EXTENSION ACTIVITY

- The children can design a menu to go with their meal.

ASSESSMENT　For assessment purposes, teachers might observe the following types of responses as children work through this unit.

at Level 3　**Designing**
- They are aware that they cannot include all the ingredients they would like, as they have to stay within the set budget.
- They explain their meal to the teacher and demonstrate that they had other ideas that they did not choose to pursue.
- They give a great deal of consideration to ensuring that their main course looks attractive.
- They decide to slice some grapes because they taste better than whole grapes and the pips could be removed.

Making
- They think ahead by preparing the main course last, as it takes longer to prepare and, if made first, would not be fresh.
- They use utensils to cut and shape the tomatoes to make them more attractive.
- Their meal looks similar to their design.

at Level 4　**Designing**
- They gather a number of photographs of meals from magazines. They use some of the ideas from the photographs in their design.
- They begin by making a list of the foods that their group think are suitable for the guest.
- They make changes at one stage because the banana has changed colour and they do not think it looks appetising.
- They design several meals from the list of ingredients that they all like. They then choose the one that most appeals to the group.

Making
- Their designs clearly show the stages that they will work through in producing the meal. All utensils and ingredients are listed.
- They shape and combine fruits and vegetables to create interesting designs.
- Their meal looks very appetising, as the food is well presented and foods have been moved and placed using tongs.
- Their evaluation clearly shows what went well and where improvements could be made if they were to make the meal again.

at Level 5 **Designing**

- Their designs take into consideration the menus they studied and the photographs of meals that they collected from magazines.
- They cut the tomatoes and place them on a paper towel to soak up the juice because they do not want the juice to run across the plate and spoil the look of the meal.
- They spend time moving and adjusting the final placing of the ingredients for the main meal. They discuss the advantages and disadvantages of placing various ingredients with others in terms of the overall appearance of the meal.
- They explain that the meal will look good and taste nice but would not be sufficiently filling for most adults. They explain what they would need to add to the meal but recognise that they are working within a budget and with a limited choice of ingredients.

Making

- They constantly refer to their plans to ensure that they undertake each task in the correct order.
- They use a wide range of utensils to change the shape and texture of the food. They choose the correct utensils from a range provided and use them safely.
- Their final meal has changed slightly from their original plans. They explain each change and give sound reasons for the changes made.

TORCHES

Context Children will evaluate a wide range of different types of torches. They will take several torches apart to gain knowledge of their basic parts. They will be provided with opportunities to match the design of the torches to their intended purpose. They will learn to make an electrical circuit.

Outcome Making a torch for a specific purpose.

Cross-curricular links
SCIENCE

Physical processes
The degree to which the following is addressed with this unit of work will depend on the age of the children, their previous experience and knowledge of electricity. It is not expected, for instance, that children must have experience of 1c. and 1d. below in order to make their torches. The degree to which the Programme of Study covering electricity is addressed is a decision that the teacher should make when planning this unit of work.

1a. That a complete circuit, including battery or power supply is needed to make electrical devices work.

1b. How switches can be used to control devices.

1c. Ways of varying the current in a circuit to make bulbs brighter or dimmer.

1d. How to represent series circuits by drawings and diagrams, and how to construct series circuits on the basis of drawings and diagrams.

RESOURCES FOR INVESTIGATING

■ A range of torches placed on display in the classroom.

RESOURCES FOR MAKING

■ A range of electrical items including:
  – 1.5V batteries
  – bulb holders
  – battery holders for 1.5V batteries
  – 2.5V and 3.5V screw-in bulbs
  – a selection of commercial switches that can be incorporated into the children's circuits
  – a selection of materials for making switches including lollipop sticks, silver foil, paper clips, drawing pins and paper fasteners
  – lengths of wire stripped at each end and the wire twisted to ease connection to bulb and battery holders
  – wire strippers
  – blocks of softwood for mounting switches on to
  – small screwdrivers.
  See Section 4 for further details regarding electrical items.

■ If opportunities are to be provided for the children to vary the brightness of their lights, a speed control switch kit will be required for inclusion in the circuits.

■ Filter papers to create coloured light (transparent sweet papers can be used).

■ Small boxes and card tubes for torch cases.

■ Silver foil and thin card for making reflectors.

ACTIVITIES   Focused practical task

■ Provide opportunities for the children to handle the range of available electrical items. Teachers may find the following ordered introduction a useful guide to planning lessons on electricity:

- Present the children with the wire strips, bulbs and batteries and ask them to light the bulb (see page 109).
- Show them how batteries can be placed in holders for ease of contact. Demonstrate the method using the bulb holder and the commercial twin battery holder.
- Ask the children to link up their circuit to two batteries and see the effect on the brightness of the bulb.
- Provide opportunities for children to investigate using one or two batteries and 2.5V and 3.5V bulbs and to make a note of the effects upon the brightness of the bulb.
- It is a good idea at this stage to link up a number of batteries to a circuit containing a 2.5V bulb. Link up the batteries one at a time and note the difference in the brightness of the bulb. Eventually the bulb will 'blow' and cannot be used again. Explain that this is what happens when too many batteries are linked to a circuit and ask the children to limit the number of batteries they use to two so that the bulbs are not 'blown'.
- If you are to use a speed control switch kit introduce it to the children at this stage.
- Ask the children to consider ways in which they can direct the light from their bulb on to a specific spot on the wall. They will begin to place shields around the bulb and some will begin to reflect the light by using a shiny object. Children can be shown how to make a simple reflector from a circle of card and silver foil as shown in Figure 3.3.

Figure 3.3

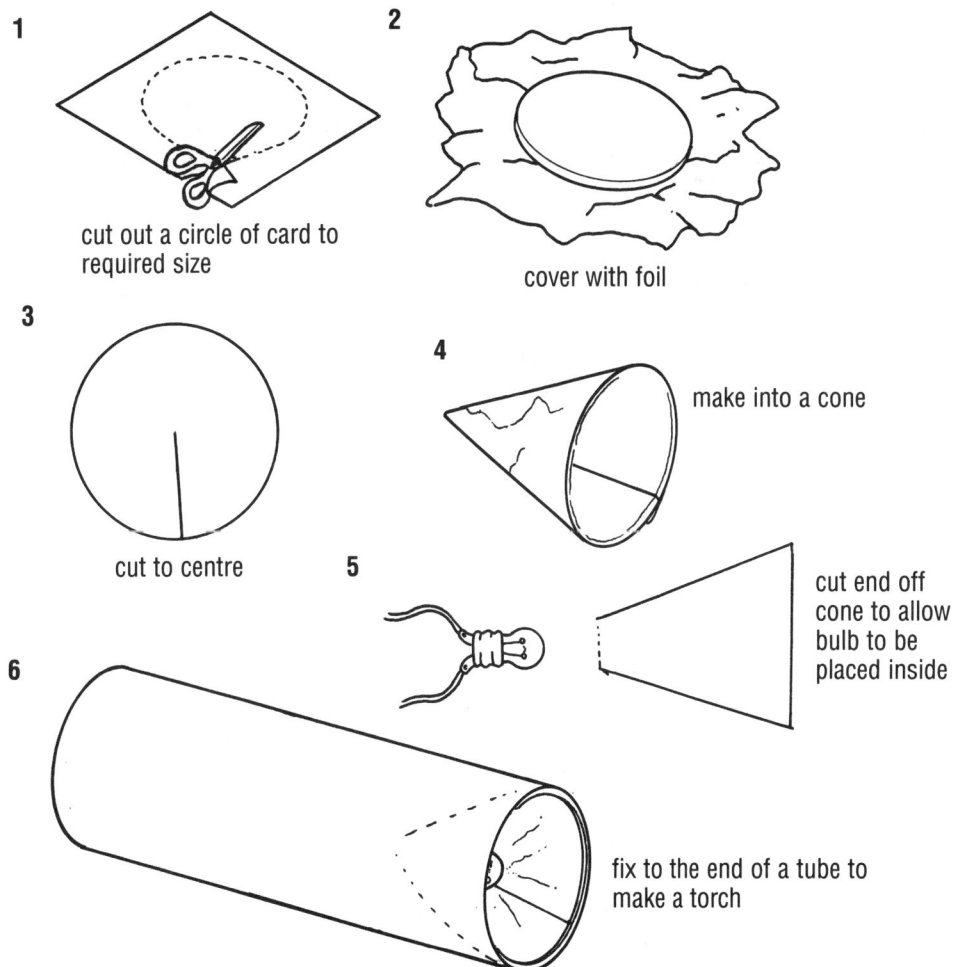

1 cut out a circle of card to required size

2 cover with foil

3 cut to centre

4 make into a cone

5 cut end off cone to allow bulb to be placed inside

6 fix to the end of a tube to make a torch

Investigative, disassembly and evaluative activities

- Gather a range of torches of different shapes and sizes. Discuss the uses for which the torches have been designed. Ask the children to explain how they know that the torch has been designed for a specific purpose by looking at its shape, the material from which it has been made, its size etc. Look at the type and size of the beam of light that is given out by directing each torch beam on to a wall from the same distance. Explain that each torch requires a light that fits the purpose for which it is to be used.

- Ask the children to choose one torch and, on a large sheet of art paper, to make a detailed drawing ot it. Encourage them to provide as much information as possible about the function of each part and the materials used, and to evaluate how well it works. Encourage them to make detailed drawings of specific parts of the torch such as the reflector or the inside view.

- Label each torch to show its specific purpose. Make a list of the children's ideas about situations in which a torch would be required. Ask the children to design a torch for a specific purpose.

- Ask the children to list the parts that all torches have. Explain that, when they come to make their own torches, they will all require a body, a light, a switch, power and a way of directing the light where required.

Design and make assignment

- Recap on the main parts to a torch that the children observed when looking at a collection of torches. Remind them that a torch can be designed for a specific purpose and that they should consider what their torches are going to be used for when designing.

- They can design and make a torch for a specific purpose, such as for:
  - looking at the back teeth in a crocodile's mouth
  - reading a book in bed at night
  - making signals or sending messages
  - an elderly or disabled person to find and use easily at night.

- Draw to the children's attention the electrical items and other materials that are to hand.

- Emphasise the need to ensure that their torches look good and are well made.

- Ask the children to design their torches and to include information about the order in which they are going to make each part, how each part is to be made and the equipment and materials that they will require.

- Discuss their ideas and, when they have thought everything through, let them begin making.

- At each stage, encourage the children to refer back to their design. Ask them to explain why they have made any changes to their original designs.

- Emphasise the importance of quality. Where children's work is not as well made or designed as it might be, ask the children to evaluate it and think of ways in which they might improve its appearance or performance.

- Provide an opportunity for the children to explain to others how they made their torches and to demonstrate how they work. Encourage the children to consider how far the torch meets the original requirements, for example, if it is for looking down plug holes, does it give off a sufficiently bright, narrow beam?

EXTENSION ACTIVITY

■ The children can design and make packaging for their torches.

ASSESSMENT

For assessment purposes, teachers might observe the following types of responses as children work through this unit.

at Level 3 **Designing**
- They design a pencil torch but realise that they can only make the body of the torch as small as the batteries. As a result, they produce designs for two other torches.
- They clearly explain how they will set about making their torch.
- Their design demonstrates a knowledge of how electrical items, including switches, can be used to achieve functional results.
- They produce labelled sketches, showing details of the circuit and how it will be fitted in to the tube they are to use.

Making
- They make the electrical circuit first and place it in the box before going on to make the reflector.
- They paint the body of the torch with Marvin mixed with paint to produce what they consider to be a good finish.
- They cut out an area of the box with accuracy so as to attach a reflector to it.
- Their torch is similar to their design. They explain that they had to attach the switch to another side of the box because it was easier.

at Level 4 **Designing**
- They collect two torches that have the same function as the torch that they are going to design. They integrate ideas from the two torches into their own designs.
- They decide to make their torch in bright colours because they say that torches need to be easy to find.
- They test the torch and discover that the beam is too large. As a result they remake the reflector so that it is smaller and narrower producing a smaller beam of light.
- They come to a decision over which they will make of the two torches that they have designed. They explain the reasoning for choosing one design rather than the other.

Making
- Their plans show a step-by-step guide to how the torch will be made. All equipment and tools to be used are clearly listed.
- They use lengths of wire passed through the body of the case and twisted on the inside to attach the switch to the torch.
- They make a simple but effective switch from drawing pins and paper clips, attach it to card and then glue it securely to the body of the torch.
- They cut the body of the torch so that the electrical items fit into it exactly. They explain that this ensures that the items do not move about inside the torch.
- They neatly stick bright pieces of metallic coloured paper to the body of the torch in a clear pattern. They explain that they designed the pattern to appeal to teenagers and young children.
- They explain that the plastic piece they have placed over the front of the torch is not ideal because it cuts out some of the light and does not provide sufficient protection to the bulb.

at Level 5 **Designing**
- They design a torch case which they cover in papier mâché in order to create a better shape to grip. They explain that the torches that they looked at were all round and could easily be dropped or knocked out of your hand.
- Before making the torch they take various tubes and check that the basic electrical equipment will fit inside and can be secured.
- They make detailed labelled sketches to explain how they are going to attach the switch to the body of the torch.
- They use paper clips and a lollipop stick with silver foil attached to produce a sliding switch.

- They evaluate their torch by reading a page from a book, using their own torch and a commercial one. They explain that their design is better because the whole of the beam is directed on to the page of the book.
- They explain that their torch could be made much slimmer by using certain types of batteries and bulbs to which they did not have access.

Making

- They work from their plans, referring to them throughout. They modify their plans when they discover that the switch they have designed cannot be attached to the Toblerone box that they chose as the body.
- They use wire strippers and small screwdrivers with confidence and take apart and reconstruct their circuits with ease.
- When making, they check through measuring that each part of their torch is the correct size so that all the parts fit together at the end.
- They produce three boxes with handles shaped from papier mâché and evaluate them before deciding which handle is the best.
- They evaluate their torches against their original designs and can explain what was successful and where changes could be made to improve the appearance or performance of the torch.

A SHOPPING BAG

Context Children will evaluate a range of bags matching properties of materials used to their intended use. They will look at the range of bags available and their uses. Children will begin to consider how different parts of bags might be made using the knowledge of textiles and the different methods of working with them that they have gained.

Outcome Making a bag from textiles for a specific purpose.

Cross-curricular links This unit can be linked to work on materials and their properties.
SCIENCE
MATHS Exploring shapes and nets of different bags.

RESOURCES FOR INVESTIGATING
■ A collection of bags on display in the classroom.

RESOURCES FOR MAKING
■ a range of fabrics suitable for making bags
■ a range of decorative materials including coloured felts, sequins and buttons
■ yarns for making handles
■ French knitting kits can be made from card tubes with four large nails taped to one end, as shown in Figure 3.4.
■ dress making pins
■ scissors suitable for cutting fabric
■ fabric crayons
■ equipment for hand sewing materials
■ a suitable, easy-to-use sewing machine if required

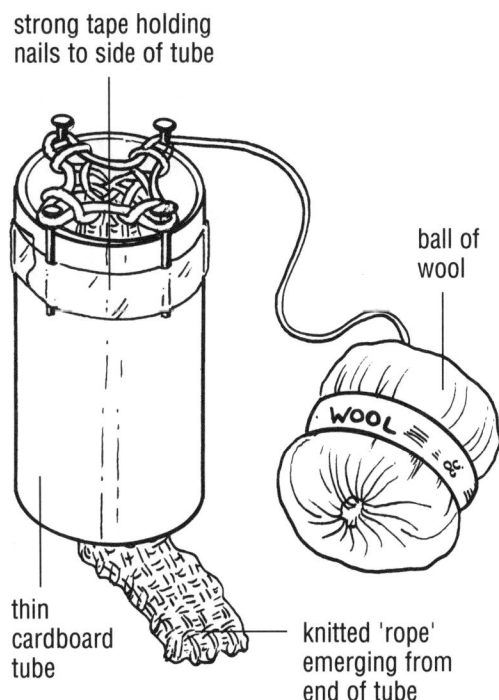

strong tape holding
nails to side of tube

ball of
wool

WOOL

thin
cardboard
tube

knitted 'rope'
emerging from
end of tube

Figure 3.4

ACTIVITIES   Focused practical task

- Demonstrate running stitch, backwards and forward stitch and blanket stitch. Provide time for the children to practise sewing prior to making their bags. Demonstrate how to sew two pieces of material together and how to add stuffing. Discuss when each stitch might be used and why. Show examples of decorative fabric work and give simple instructions.
- Demonstrate to the children different ways of making bag handles from yarn by plaiting, weaving, French knitting and simply twisting the cords. Provide time for the children to practise the techniques learnt.
- Show the children different ways of attaching handles, including loops, staples, hand stitching and machine stitching (if taught in the school).
- Provide opportunities for the children to explore the properties of the different fabrics. Simple stretching tests or water tests (to see which is most waterproof) can be carried out on each fabric.
- Provide good hand-held and stand-alone magnifiers and binocular microscopes with which to observe the patterns of the weaves of different materials.
- Demonstrate the use of fabric crayons.
- Allow children to explore how designs will look on different types of fabric.
- Show the children how to cut out their paper pattern, pin it to the fabric, cut out the pattern and sew it together. Allow them to practise the technique if necessary, by pinning and cutting out patterns using scrap pieces of fabric.

Investigative, disassembly and evaluative activities

- Ask the children to help you to make a collection of bags. Display them in the classroom.
- Ask the children to choose a bag and consider what the bag might be used for. Ask the children to make a detailed drawing of their bag on a large piece of drawing paper. Ask them to provide as much information as possible, including the materials it is made from, how they think certain parts were made, which bits they like best, who they think the bag was made for and how well made they think it is. The children may give any other information that they discover about their bag, for example, where it was made.
- Provide a bag that children can take to pieces and explore how it was made and the materials used in its construction.
- Provide opportunities for the children to explain to others what they have discovered about their bag.

Design and make assignment

- Remind the children about the techniques they have learned and about safe ways of working.
- Show the children all the materials and equipment available for making the bags.
- Ask the children to begin to design their bags. They should first consider who the bag is for and what it will be used for.
- Ask the children to make sketches and to show as much information as possible, including a drawing of their bag, the stages they will work through in making the bag and all the materials and equipment they intend to use.
- Encourage the children to model their bags with sugar paper and strips of card that can be stapled or stuck together.
- Once the children have thought through how they will make their bags, they can begin making.

- Encourage the children to consult their plans throughout. At regular intervals, stop the children and ask them to explain how they are proceeding and any alterations they have made to their original plans.
- Emphasise the need for quality in their making. Whenever you think the quality is not as good as it might be, draw this to their attention.
- Encourage the children to explain to others how they made their bags, the ways in which they meet the original designs and ways in which they could be improved.

EXTENSION ACTIVITY

The children could create a full-page advert for their bags in a Sunday magazine.

ASSESSMENT

For assessment purposes, teachers might observe the following types of responses as children work through this unit.

Designing

at Level 3
- They have several ideas for bags. They explain that one bag that they all like will take too long to make.
- They explain how they will set about cutting out and joining the two pieces of material.
- They choose a material because it is stronger than the others and will therefore not wear out if, for example, groceries are carried in it.

Making

- They plan the stages for making their bag.
- They choose the most appropriate tools for sewing the two pieces of material together using a simple running stitch.
- They work with some degree of accuracy in making the bag.
- Their bags look similar to their original design.

Designing

at Level 4
- They decide to make a shoulder bag for someone their own age. They collect a few bags together and take the best ideas to use in their own designs.
- They ask members of their class to choose what they consider to be the best design and give reasons. They use the feedback to inform their final designs.
- As it develops, they check their work against their design.
- Their designs show a number of ideas, all of which are clearly labelled. They set down points for and against each design.

Making

- They set down the stages they will need to work through in order to produce their bag. They set down all the tools and materials that they are to use.
- Before they set about joining their two pieces of material together, they talk through the stages they will have to work through to ensure they have thought everything through.
- They cut and join the materials, having first made a pattern from paper and pinned it to the material. They are aware of the different stitches they can use to attach decorative objects to the bag.
- Because they have made careful checks at each stage, the finished bag is the exact size set down on their design.

Designing

at Level 5
- They test all the materials and find one that is least affected by being placed in water over a period of time. They decide to use this material to make a bum bag for swimmers.
- They are unsure if their method for joining the handle to their bag will work so they test out the idea on a piece of material.
- They model their bag and use a safety pin to attach a handle made from string. One child wears the model bag whilst the others change the size of the handle until they feel it is just right.

- When wearing the model bag, they note that it crumples up and loses its shape. They decide that they will sew together two pieces of material at the back and stuff it in order to ensure that the bag keeps its shape.

Making

- They work from their plans throughout but change the design of the fastener, as they realise it would be too difficult to sew it on firmly without a sewing machine.
- They use sewing equipment safely and with confidence and make their own patterns and use them to sew together the sides of their bag.
- They sew two pieces of material together using a backwards and forwards stitch, which they explained was a stronger stitch and provided a better finish when the bag was turned inside out. They use blanket stitching to decorate the open top of their bag.
- They note that their bag is still crumpling up when worn so they stick a thick piece of card across the opening to provide rigidity. They cover the card neatly so that it does not spoil the look of the bag.
- They evaluate their bag and explain how it might have been improved. Overall, they think they have made a very good bag and can explain why.

READING BOOKS

Context	The children will evaluate a range of reading books for young children to identify their main features. They will identify books that they enjoyed and which young children in the school prefer. They will develop a knowledge of what young children find attractive to look at and have read to them.
Outcome	Making a reading book that can be read to a child in the reception class.
Cross-curricular links **ENGLISH**	Reading and writing. Speaking and listening – reading to an audience.

RESOURCES FOR INVESTIGATING	■ A selection of reading books from the reception class. Try to choose a set of books that are different shapes, contain interesting pictures, have limited text and are extremely popular (e.g. *The Hungry Caterpillar*). Also include books that the class remember from their own experience. You might decide to include books with pop-up or moving sections to encourage the children to produce similar effects. If children plan to make such books, additional time will be required for this activity.
RESOURCES FOR MAKING	■ Card of various thicknesses, colours and sizes. ■ Where available, a full computer system plus colour printer. Some children may decide to design covers using a graphics program or clip art. A word processor can be used to produce text in various styles and sizes.

ACTIVITIES

Focused practical task

■ Decide on the methods that the children could use to make their books. There should be a choice of methods but not too many, or the making will be difficult to manage in the classroom situation. Possible ways of making books include:
  – folding a sheet of thin card in half and adding the required number of sheets of the type of paper you require, stapled to the inside

Figure 3.5

children can make their own books using a long-arm stapler

Figure 3.6

– joining several long strips of thin card and folding concertina style. This method can be used for fold-out and stand-up books, which are good for display. If displayed, the books are much more likely to attract children to read them. Both sides of the book can be used if required

– making a sewn and bound book

– using a spiral binder with card covers front and back. Plastic spiral binders can be cut in half using a junior hacksaw to produce A5 size.

There are many other methods and the children will come across ideas, either from the collection of books provided or from books of their own.

Figures 3.7 and 3.8

mark the four points down the centre of the book

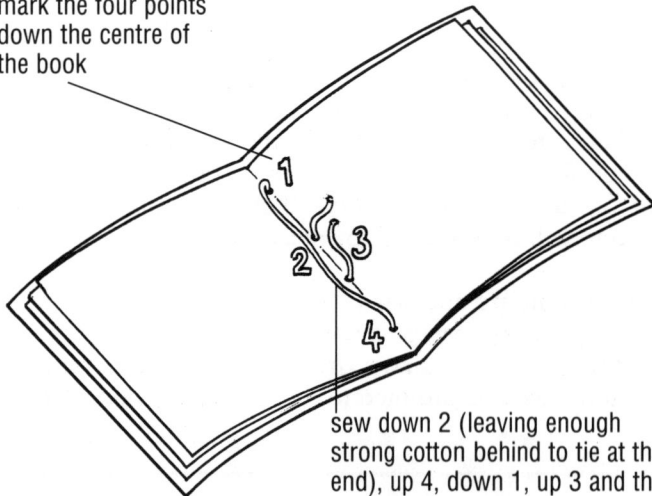

sew down 2 (leaving enough strong cotton behind to tie at the end), up 4, down 1, up 3 and then tie the ends together

use a book binder to make ringbound books

■ Demonstrate the methods you are going to use and provide an opportunity for the children to practise the techniques by making a simple book for the next project you undertake.

■ Show the children how to use the strimmer safely.

■ Demonstrate a number of techniques for printing patterns on to paper. Children could practise by producing covers for their project books.

■ Demonstrate how to create drawings and patterns on a computer graphic package and how to print them out in colour. Flare using the Archimedes system, or Colour Magic on the RM system, are particularly good programs for producing simple patterns which can be adapted, making a symmetrical or tiled pattern which can be printed out and used as book covers. Simple instructions could be written out and placed next to the computer system for the children to refer to throughout the project. If some of the children are particularly good at using the computer, set them the task of explaining the method to less able children. Emphasise that they should not do the work for them but help them to do it themselves.

■ Make a simple push, pull or slide mechanism from paper and card (see page 113). Show it to the children and ask them to explain how they think you made it. Ask them how it might be used in a book. Ask them to make their own moving parts from paper and card using the techniques from your example and their own ideas. Set them the challenge of making mechanisms that slide, push, pull, twist, close etc.

Investigative, disassembly and evaluative activities

- Ask the children what makes a good book to read to a reception child. Make a list of their ideas, which should include the cover, the story and the illustrations.
- Ask them to look at the books you have collected and to evaluate them against the list of ideas that you have just drawn up. Explain that the books are from the reception class. Provide an opportunity for members of the class to talk through their evaluations. They could choose what they thought was the best book and explain why.
- Make a graph to show which of the books you have collected the class feel are the best for young children. The books can be arranged in order of popularity to create a 3D 'graph' in the classroom.
- Provide an opportunity for the children to read their chosen book to a reception child. They should evaluate whether the reception child thought the book was as good as they thought it was. Why did the young child like or not like the book?

Design and make assignment

- Discuss with the children all the things they have found out about what makes a good book for reading to a young child. Explain that they are to design a book that they are going to read to one of the reception class or to a group of reception children.
- Show them the equipment and materials that are available and remind them of the need to set up a working area and to work safely at all times. Remind them of the techniques that they have learnt when working with the materials available.
- Ask them to record their designs, which should show what the finished book will look like, including the layout for each page. Their designs should include the stages in which they will make each part of the book and a list of all the equipment and materials that they will require.
- Remind the children that the book should look good and great care should be taken over every stage of making their book. Bring to the children's attention any aspect of the making of the book that might require extra care if quality is to be maintained. This may include the text where children with untidy handwriting may choose to use a word processor.
- Throughout the making stage, encourage the children to evaluate their progress and to refer to their designs.
- The children could take their completed books and read them to the reception children. They should use the opportunity to ask the young children to evaluate the books.

EXTENSION ACTIVITY

- Provide guidance to the children on ways in which they can make their stories more interesting to the young children. This work links well with English, speaking and listening, providing opportunities for children to read to an audience. Show them how to emphasise parts of the story, how to change the pitch of their voice, use their eyes and to bring the pictures or moving parts to the attention of the listener to make the story more interesting. Allow the children to practise reading their stories before they visit the reception class.

ASSESSMENT For assessment purposes, teachers might observe the following types of responses as children work through this unit.

at Level 3 **Designing**
- They design a book that will stand up concertina style but explain that there is a lot of waste because they will not be able to write anything on the back of each page. However, they decide to make it because they say it is going to be much more attractive to the reception children than the others they have designed.
- They explain how to make the book and produce the pages on the word processor.
- They design a simple pop-up character for the front of the book, to be made from card with a fabric face attached.
- They produce a design for each page of the book. The front cover design is neatly labelled showing the materials to be used to produce the pop-up character.

Making
- They make the pop-up character first, then make the front cover, before attaching the figure and checking it works correctly.
- They precisely cut out the text produced on the word processor, using the strimmer correctly. They arrange the text on the screen so that it fits around the drawings they have made in their books.
- Their finished book looks similar to their original design except for the overall size, which had to be reduced because it would have required more than a single sheet of card, involving waste.

at Level 4 **Designing**
- They talk to the reception children to discover which are their favourite stories. They use the information to help them decide on the type of story to write.
- They produce several ideas for a book and explain the good and bad points of each
- They consult the younger children and ask them which drawings they like best.

Making
- They produce step-by-step designs showing how the book is to be produced and listing all the materials and tools required.
- They record the making process for each page, linking drawings and text on a computer.
- They cut and arrange the moving parts for each page with care and accuracy.
- They reinforce all their pop-up characters because they feel that their original ones would soon be broken by young children.

at Level 5 **Designing**
- Their designs take account of ideas from books collected from the reception class.
- They model the moving parts for each page from paper and card to ensure that their ideas will work.
- They plan to cover the electrical wires that are attached to a buzzer placed on the front of the book. They incorporate the small single battery into a 3D design for a house that is attached to the front cover.
- They have set down in their plans a number of materials that can create the textures that the young children can feel on each page.
- The story, drawings and tasks on each page are well matched to the age of children for which the book is designed.

Making
- They work from their plans throughout the making stages.
- They attach the moving parts with precision and take considerable care to ensure that each page is of the best possible quality. Everything is mounted with care and loose parts are 'hidden' away.
- They write several short stories and read them to a group of children from the reception class, asking for their views on which is the best.
- They read their book to a group of children from the reception class and explain ways in which they could have improved the book, based upon their own evaluation and that of the younger children.

SOAP DISHES

Context Children will look at a range of soap dishes and evaluate their designs. They will look at different soaps and match them to the designs of the soap dishes they have collected. They may observe a clay dish being made at a potters so that they can draw upon the ideas they observe when designing and making their own dish.

Outcome Making a soap dish from clay or other suitable mouldable material.

Cross-curricular links
HISTORY This project can be linked with work on the Victorians if a suitable collection of antique soap dishes can be collected. Links can be made between designs then and now with older and newer soap dishes placed on display.

RESOURCES FOR
INVESTIGATING
■ A selection of soaps of different sizes and shapes.
■ An example of two very different soap dishes.

RESOURCES FOR
MAKING
■ Tools for moulding and shaping clay
■ A kiln, if available, but this is not essential as clay can be left to dry and painted with either water-based glue mixed with a small amount of paint, or readymix paints.

ACTIVITIES   Focused practical task

■ Demonstrate how to work the clay and techniques for making particular shapes. Show the children how to make thumb pots, coil pots and, if the equipment is available, how to use moulds. Provide opportunities for the children to practise and develop their skills at working with clay.
■ Emphasise how to set out the working area so that everything is to hand. The children should be developing greater responsibility for setting out and clearing up their own working area – particularly important when working with clay.
■ If a kiln is available, several pots should be fired to show the children how this is done and how much care is required in making a clay pot if it is not to break up when fired.
■ If a kiln is not available, the children should be shown how to mix paints and what pots look like after they have been painted. If possible, show them examples of clay which has been fired, even though they will not be able to use this method.

Investigative, disassembly and evaluative activities

- Make a collection of soaps and draw to the children's attention the different shapes, sizes, smells and colours. Allow the children to handle the soaps. Ask them to explain any problems that they have at home with soap. It is highly likely that they will mention the following:
  - the soap goes slimy and leaves a mess by the sink
  - there is nowhere to place the soap, and it keeps sliding down into the sink.
- Display various soap dishes and ask the children to make labelled sketches of one, showing any interesting features, what it is made from, where it was made and any other information that they think will be of interest to others.
- One of the soap dishes may have a hole in the base, through which water can drain to stop the soap from going soft; others might have more than one space so that another item can be stored next to the soap.
- Let the children try placing the soaps into the dishes and evaluate how well each dish is designed to take particular soaps. Ask them to consider where the particular soap dish would be placed – they may be designed to be placed next to a sink, on a sink, in a shower as a tile fitting, overhanging a bath or they may be antique and designed to fit at the side of a flat-topped sink.

Design and make assignment

- Ask the children to choose one of the soaps for which they will design a soap dish. Tell them that they are going to design and make a soap dish that is to be given as a free gift with every bar of the soap or as a present to a particular person. Ask them to tell you what they need to consider when they are designing their dishes. They may suggest:
  - the size of the soap in relation to the dish
  - how attractive the soap looks in the dish bearing in mind colour and shape
  - where the dish is going to be sited.
- Ask the children to make a sketch of the dish they are going to make and label it carefully to show any detail or design. They should list all the tools and materials that they will need to make their dish. They will also need to give careful thought to the size of the dish because it is going to be used with a particular bar of soap.
- Once the children have thought through each stage of the making, they can begin to set out their working area.
- Encourage the children to consult their designs regularly to guide them in their making. Ask them to explain why they have made any changes to their designs.
- The children can place their completed dishes on display, along with the bar of soap. They should be provided with opportunities to evaluate their dishes against their original designs and to explain to others what they think worked well and what they would do to improve the dishes if they were to make them again.

EXTENSION ACTIVITIES

- They can make their own 'soap' shape from a suitable mouldable material. Some young children may find several of the commercial bars of soap too large for their hands and will enjoy making their own soap shape and then designing and making a soap dish for it. Listed in Section 4 under 'Mouldable materials' (see pages 106–107) are recipes for mouldable dough. The children can not only design their own shaped soap but they can add a scent as well.

■ They can design a package for a special offer of a free soap dish with every bar of soap.

ASSESSMENT For assessment purposes, teachers might observe the following types of responses as children work through this unit.

at Level 3 Designing
- They make several different designs for soap dishes and, although they all have one favourite, they recognise that the cover made from clay looks good but could be easily broken.
- They explain how they will make their dish from clay.
- In their designing they draw upon the ideas they gained from looking at the collection of soap dishes.
- They produce labelled sketches.

Making
- They list the tools and materials to be used in their designs.
- They use the shaping tools accurately and decide to varnish the finished dishes to give a more attractive, glossy look.
- They cut and shape the clay to make their dishes.
- Their dishes are similar to their designs.

at Level 4 Designing
- They find a book in the school library on the Victorians which shows examples of soap dishes. They use some of the ideas within their own designs.
- They ask their parents what they would like and, as a result, they decide that their dish should fit between the taps of a sink.
- They reject several ideas because they would easily break, were too big for the soap or, when they checked, would not fit between the taps as they had intended.
- They have several designs for dishes and spend some time discussing the advantages and disadvantages of each before making a final choice of which to make.

Making
- Their designs include step-by-step plans for each stage of making and list the tools and materials required at each stage.
- They demonstrate a knowledge of a range of techniques for making their dishes from clay. They decide on the most appropriate technique, from a range, for making their dishes.
- They check throughout that the dish is the correct size and that their soap makes a good fit.
- They decide to use paints that match the wrapper on the soap they are using. They paint several varieties of colours on a large sheet of paper and then place the soap alongside to see which colours are the best match.
- They evaluate their soap dishes, pointing out what are their good points and ways in which they might improve them.

at Level 5 Designing
- Their designs draw upon ideas gained from a visit to a potter, who demonstrated how to make a dish. They include patterns around their dish which they explain they are going to make in the same way as the potter did.
- They make a small version of their soap dish from Plasticene in order to check that water will run through the hole they have made into it. They realise that the underneath of their soap dish needs to be raised, so that the water can drain away.
- They decide that the soap dish should be simple and should not use too much clay, otherwise it will cost the soap company so much money to have them made that it will make no profits.
- Their final design uses the least amount of clay possible and would be the easiest to make.

Making

- They work from their plans throughout but adapt them when they realise that the colour they have chosen clashes with the colour of the wrapper on the soap.
- They use the tools for cutting and shaping the clay with confidence and accuracy.
- They check the size and shape of their dishes throughout, ensuring that the dimensions are almost identical to those on their plans.
- They evaluate their dishes against the criteria set for designing. They explain how they might improve the dishes if they were to make them again.

LIFTING AND LOWERING

Context Children will develop a knowledge of the range of lifting and lowering machines and the objects that they transport. They will observe, firsthand, machines. From their observations they will begin to develop a clearer understanding of the ways in which machines lift and lower objects. They will explore movements through working with construction kits and applying what they learn to real life applications.

Outcome Making a lifting and lowering machine.

Cross-curricular links
SCIENCE This unit can be linked with work on forces and motion.

RESOURCES FOR INVESTIGATING
- Toys that lift and lower and any photographs showing a variety of machines, especially cranes and tipper trucks.
- Books and other reference materials illustrating machines and especially those that explain how they work.
- Lego Technic kits 1 and 2 and accompanying activity cards.
- Pneumatics and hydraulics kits – see page 112.

RESOURCES FOR MAKING
- corrugated sheets
- fasteners
- wood strips
- dowel
- wooden wheels
- any materials suitable for making or using as pulleys – see pages 114–115

ACTIVITIES

Focused practical task

- Provide opportunities for children to apply the knowledge and skills developed at the investigative, disassembly and evaluative stage. Provide them with a small weight and ask them to make a device that will slowly lower and lift it from the table top to the floor.
- Ask the children to mount a pneumatics system that will turn a switch on or move an object so that it falls off the end of the table.
- As the children work, encourage them to evaluate their mechanisms and look for ways in which they can be improved. Ask them to consider ways in which their mechanism can be secured or weighted.
- Set up a display of examples of the mechanisms children have made and encourage them to relate them to ways in which they can be applied in machines that lift and lower.
- Demonstrate to the children the use of gears using a suitable construction kit such as Lego Technic or Start Gear. Show how changing the gears can speed up or slow down the mechanism.

Guidance for teachers

Changing the speed at which something rotates is achieved by placing together gears which have different numbers of teeth. It is important to point out to young children that gears have teeth and to explore the effects of meshing together gears with different numbers of teeth. Children can use Start Gear or gears from a Lego Technic kit to explore gear ratios. If, for instance, they use the gears from the Lego Technic kits then they can take a gear with forty teeth and position it so that when turned it rotates another gear with eight teeth. They will immediately be able to observe that the rotation has been speeded up so that the gear with eight teeth rotates more quickly than the gear with forty teeth. The number of turns from one gear to another is known as the gear ratio. The gear ratio in this particular example would be written as 1:5. Point out that, for every one turn of the large gear, the small gear will rotate five times.

Investigative, disassembly and evaluative activities

- Provide the children with the Lego Technic kits and any of the activity cards which relate to lowering and lifting. Ask the children to make the mechanisms shown on the activity cards. Discuss how their mechanisms work and where they might be used. The activity card provides some ideas for applications of the mechanisms that the children build.
- Provide the children with the pneumatic and hydraulic kits and materials and show how they can be used. Provide the correct name for each item, explain how it can be used and ensure that the children use the correct terms when describing their mechanisms to others. If you do not have access to water or conditions are not suitable for using water, introduce the children to pneumatics. Provide objects such as a brick or chair which they can push along with their pneumatic system. The children should investigate the difference between using small and large syringes and using water rather than air. Which provides the most powerful push?
- Set the children the challenge of using one syringe to push out and pull in two other syringes.
- Ask the children to explain their mechanisms to others and, wherever possible, ensure the children use the correct vocabulary, such as gears, pulleys, gearing ratios, pneumatics, hydraulics.

Design and make assignment

- Recap on all the work that has been covered and the different methods that can be used to make a lifting and lowering machine. Show the children the range of materials available to them.
- Ask the children to design a machine for lifting, or lowering, or lifting and lowering. They should consider what they are going to lift and lower and whether it needs to be transported from one place to another.
- The children need to provide considerable detail in their designs so that it is clear what they are going to make and how they will make it.
- Children can set down their designs in a flow format to show the stages that they will work through when making.
- Encourage the children to think of alternative ideas at each stage so that they can evaluate their ideas and have other options open to them if, at the making stage, they discover that things have not worked out as expected.
- Encourage the children to consider how their machine will look. Can they think of ways of hiding away the mechanisms so that in real situations people would not be injured?
- Once the children are clear about what they are trying to achieve, they can begin making.
- Stop the children at certain stages and ask them to explain to you and others the progress they have made. Encourage children to use appropriate vocabulary when describing their machines.

■ Provide opportunities for the children to demonstrate their completed machines and explain how they were made. They should be given time to evaluate their machines and to say how they would improve them if they were to make them again.

EXTENSION ACTIVITIES

ASSESSMENT

■ Children can make a vehicle into which their lifting machine can deposit its load.
■ Children can add lights to their vehicles, if appropriate.

For assessment purposes, teachers might observe the following types of responses as children work through this unit.

at Level 3 **Designing**
• They generate a number of ideas for machines to make.
• They select one design because, although the mechanism that they have made from Lego does not work as well as others, it is small enough to fit on to the mobile wooden chassis.
• Their designs clearly show their understanding of the use of pulleys and gears to create different forms of movement.
• They produce labelled sketches.

Making
• They make a Lego mechanism first because they know that its size will dictate the size of the frame or chassis that they have to make for it.
• They use tools to cut and shape corrugated coloured card which they make into a simple chassis on which to mount their Lego mechanism.
• Their lifting and lowering mechanism looks similar to their design.
• They explain changes they have made to their Lego mechanism in order to make it lift things more quickly.

at Level 4 **Designing**
• They decide that they want to create a simple lifting and lowering mechanism that can be mounted on to a vehicle. They gather the appropriate activity cards and construct a number of the examples given. They choose, from the ones they have made, the one they think will be the most appropriate.
• They choose corrugated card that will match the colour of the Lego mechanism that they have made. They place signs on their vehicle and take great care over how it looks.
• They constantly test out their mechanism, which has to lift articles on to the rear of a vehicle for transportation.

Making
• They produce step-by-step plans showing each stage of making.
• They list all tools and materials to be used. Their designs show how the rear flap will be produced using coloured corrugated sheets.
• Their mechanism is well finished and works as required. Overall, a great deal of care has been given to how the mechanism looks and works.
• They identify what is working well and what might be improved.

at Level 5 **Designing**
• Their designs demonstrate their understanding of how gears and pulleys are used to create different forms of movement. They explain that certain of their ideas come from vehicles they observed on their visit to a building site.
• They model a number of mechanisms from Lego Technic kits, showing understanding of how the various parts fit together to create different forms of movement.
• They use corrugated sheets and appropriate fasteners to create frames for vehicles or in which to house their mechanism.
• They evaluate each of their designs, explaining the advantages and disadvantages of making each, in relation to its function.

Making

- They work from their plans but modify them when they have difficulty making their mechanism balance and have to add a counter balance.
- They use Lego Technic kits, corrugated card and wooden strips to produce mechanisms and frameworks with confidence. They choose the most appropriate material and technique after evaluating the options available at each stage.
- They evaluate their making against their design at each stage.
- When evaluating their final mechanism they evaluate it against their original design and its intended function.

TOOLS AND MATERIALS

Equipment required for the Scheme of Work

This section provides a list of tools and materials that might be required for undertaking design and technology activities. This is not a definitive list of all the equipment and materials that primary schools should purchase; many schools will find that they have alternative equipment that they use or, indeed, have far more resources than those listed. The list gives details of the tools and materials that schools will require to undertake the units of work set out in Section 3, 'A Scheme of Work'.

The majority of items listed can be purchased from Technology Teaching Systems (T.T.S.) Chesterfield or G.C. Products, Scredington, Lincolnshire. Addresses of these and other suppliers can be found in the Useful addresses section on page 117.

SHEET MATERIALS

- A wide variety of card of different colours and thicknesses.
- Appropriate tools for scoring and cutting card.
- Corrugated plastic sheet – which is supplied in different thicknesses (3mm and 4mm being the most common) and colours.

Equipment for use with sheet materials

- joiners
- click rivets
- suitable tape
- hole punch
- 5 mm dowel for pushing into the sheets to create axles
- small and large strimmers that are safe for children to use
- safety snips – these are useful for cutting reclaimed materials, thick card, plastic etc.
- scissors

RECLAIMED MATERIALS

Reclaimed materials can be difficult to store so it is advisable to sort the materials you require. Store the materials in large, labelled boxes so that children can take responsibility for ensuring that they are regularly topped-up with items brought from home. Ensure that children are aware that they should only bring clean items into school.

The ways of categorising the boxes will depend a great deal upon the ages of the children who are collecting the items. With older children the collection can be linked with work in science on classification of materials.

A suggested way of sorting might be:

Tubes
- card tubes (toilet rolls should not be used)

Boxes
- boxes of different sizes (they can be covered in paper to make painting easier *or* opened and taped together inside out)

Plastic items
- yoghurt pots
- margarine tubs
- ice cream tubs

Clear and shiny items
- clear plastic
- aluminium foil
- foil dishes

Wooden items
- small wooden boxes
- odd wooden strips and blocks
- purchased lollipop sticks (used lollipop sticks should not be used for health reasons)

FRAMEWORK MATERIALS

Wooden strips and card triangles

Joining wood strips using card triangles is a useful method for constructing frameworks. However, it requires careful planning beforehand when precise measurements have to be worked out for each wooden strip. The strips have to be measured accurately prior to cutting in order to reduce waste.

- 10 mm x 10 mm wooden strips
- 5 mm dowel
- wooden wheels with pre-cut centres to fit 5 mm dowel
- cutting and sticking boards
- pre-cut card triangles
- clear PVC tubing for spacers on axles – 5 mm inside diameter
- PVA glue
- glue guns to speed up the process of construction. Teachers must consult health and safety regulations regarding the use of glue guns in schools.

Equipment for use with wooden strips and card triangles

Figure 4.1 shows a technique for making frameworks from wooden strips and card triangles.

- junior hacksaws
- hand drills
- drill bit for creating tight fit on dowel
- drill bit for creating loose fit on dowel
- vice to grip desk or other work surface – place piece of wood underneath to ensure the work surface is not marked
- clear plastic jig for guiding cutting and drilling – sometimes called Richardson block
- G-clamps
- safety rule
- cutting and sticking boards
- lynx joiner for accurate joints
- sanding sticks
- single hole punch

Other framework techniques

There are a wide range of techniques and materials for creating frameworks, including:
- construction kits
- rolling paper into tubes and joining together
- making paper into square tubes and joining in a variety of ways

– cutting card into strips, which can form frameworks within which mechanisms can fit, e.g. when creating moving parts on a greeting card (see also 'Mechanical components').

Figure 4.1

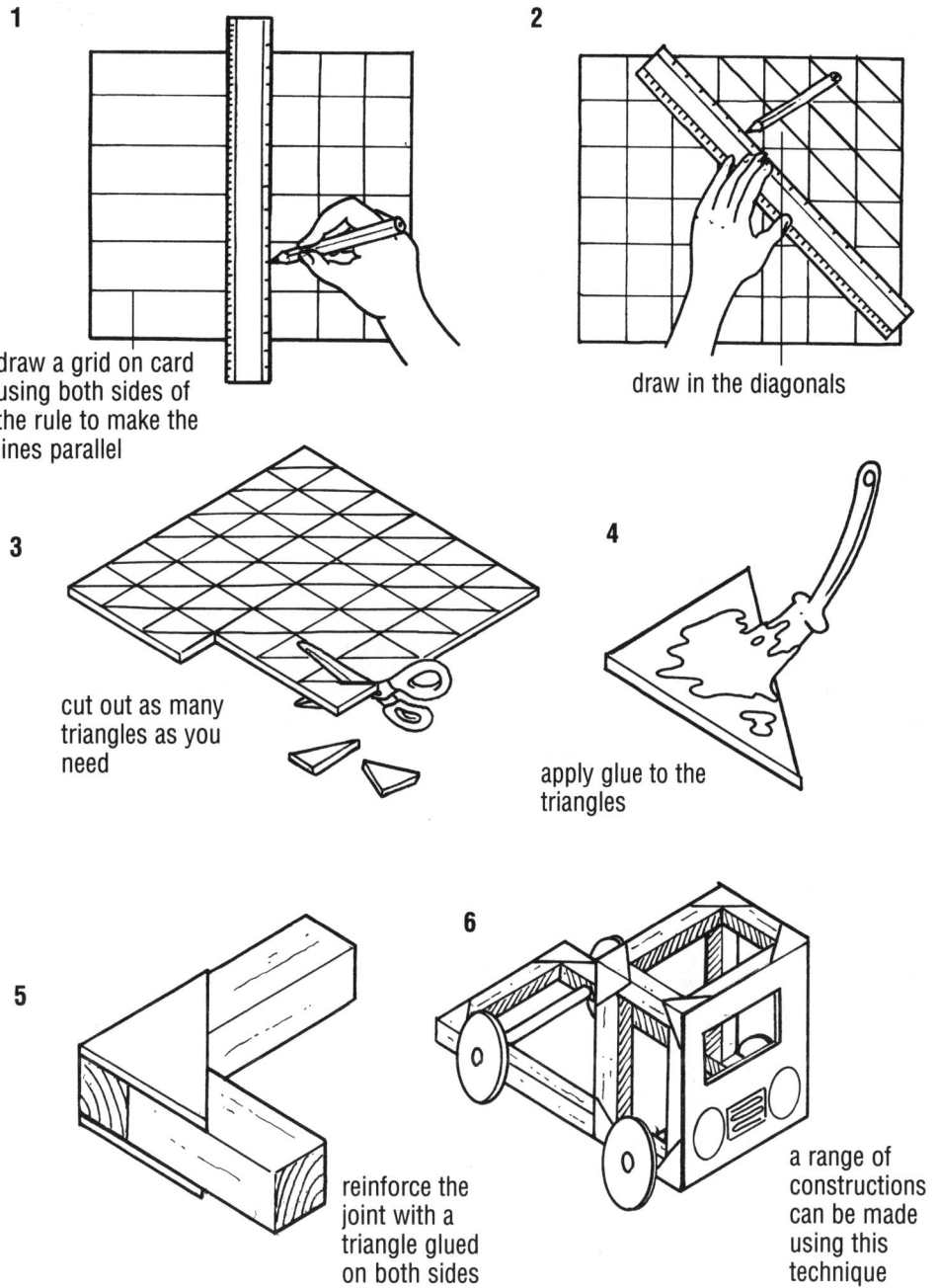

1 draw a grid on card using both sides of the rule to make the lines parallel

2 draw in the diagonals

3 cut out as many triangles as you need

4 apply glue to the triangles

5 reinforce the joint with a triangle glued on both sides

6 a range of constructions can be made using this technique

MOULDABLE MATERIALS

- clay – choose the most suitable for your particular use and the age of the children
- papier mâché
- Plasticene
- playdough

Home-made doughs There are a range of doughs which teachers can make from basic ingredients to match their needs. Listed below are a number of the most popular recipes:

Basic dough mix (1)
Mix together:
½ cup salt
1 ½ cups plain flour
water

Scented dough
½ cup salt
1 ½ cups plain flour
water
food colouring

Mix together adding more water until a mouldable mix is produced. Then add a few drops of vanilla, peppermint or almond essence.
Bag and then store in an airtight plastic tub.

Craft clay
½ cup cornflour
1 cup bicarbonate of soda
½ cup water

Combine ingredients in a saucepan and cook over medium heat until of a dough-like consistency.
Turn on to a board and knead well.
Cover with a damp cloth until cool and then cover with tin foil.

Basic dough mix (2)
Mix together:
½ cup salt
1 ½ cups self raising flour
water
food colouring

Warmed dough
1 cup plain flour
½ cup salt
1 tablespoon cooking oil
1 tablespoon cream of tartar
1 cup water
food colouring

Warm all ingredients in a saucepan.
This dough will keep for half a term if placed in plastic in a zip-lock bag then stored in an airtight plastic tub.

Modelling dough

1 cup salt
¼ cup water
1 cup cornflour

Mix salt and water together in saucepan. Stir constantly over a medium heat for 3–4 minutes until the mixture bubbles.
Remove from the heat and immediately stir in the cornflour and a further ½ cup of water. Stir until stiff.
Knead well until pliable.
Store in an airtight container. Will harden if exposed to air for thirty-six hours or can be baked in the oven at 360° F.
Can be painted when dry.

Snow dough

2 cups laundry soap flakes
¼ cup water

Whip together with a hand whisk, adding more soap or water as needed until the mix resembles thick whipped cream.
Dip hands in water before moulding so hands do not get the dough stuck on them.
Dries to a porous texture that lasts for weeks. Best kept overnight after mixing and before modelling.
Can be used for a snow effect or for frosting card.

Health and Safety   Many detergents contain ingredients which can irritate the skin and eyes. Read the instructions on the box for any health warnings. Children should wear rubber gloves unless teachers are sure the soap is safe.

TEXTILES

Fabrics
- felt
- calico
- hessian
- crimplene
- acrylic fleece
- cotton-type fabrics
- white cotton or muslin to use with fabric crayons

Threads
- acrylic yarn
- Sylko perle cotton No. 5 or 8

Glues
- PVA glue for sticking fabric to fabric
- diluted PVA glue for sticking fabric to paper
- velcro strips

Needles
- tapestry needles No. 16 or 24 for use on open fabrics
- chenille needles No. 14 to 18 for use on close fabrics

Please note, the larger the number, the thinner the needle

Dyes and paints
- felt pens
- Pentel pastel dye sticks
- Dylon fabric painting pens
- Colourfun paints
- fabric crayons

It is essential that all dyes and paints used are non-toxic. Children should wear aprons or protective clothing when using materials which can stain fabric.

Other items
- dressmaking pins
- stuffing (e.g. kapok)
- decorative materials including:
  – buttons
  – coloured net
  – sequins
  – felt offcuts
  – coloured lace

Food

The list below is based on the assumption that schools do not have access to cooking facilities. Refer to relevant Health and Safety Policies regarding the purchase and use of suitable utensils.

Figure 4.2

- weighing scales
- measuring jugs
- plates
- saucers
- plastic knives
- plastic forks
- plastic spoons
- clear plastic cups
- tablespoons
- dessert spoons
- spatulas
- sieves
- mixing bowls
- chopping boards
- cutters
- shapers
- stamps (for cutting)
- rolling pins
- wooden spoons
- measuring spoons
- peelers
- graters
- crinkle cutters
- mashers
- colanders
- lemon squeezers
- fish slices
- PVC sheeting for preparation area (specifically for use when working with food)
- suitable sterilising agent (e.g. Milton)
- washing-up bowls
- tea towels
- dishcloths
- aprons specifically for use when working with food
- kitchen roll
- aluminium foil

schools may choose from the range of food toolboards available to assist with the safe and hygienic storage of equipment

CONSTRUCTION KITS

- Knex
- Small Duplo
- Lego Duplo kits, especially kits that focus upon simple mechanisms and forms of movement
- Lego Technic kits plus activity cards (N.B. Lego Technic 2 contains a 4.5V motor and battery holder)
- Start Gear
- Clixi
- Reo Click
- Large construction kits such as:
  - **Quadro** – a larger version of Mini-Quadro. Large constructions can be made including vehicles, shops or puppet theatres. However, it does require sufficient storage and work space. Purchased from Heron Educational.
  - **Lasy** – comes in small, medium and large kit form. Wheels, motors and gears are available to be incorporated into designs. Purchased from Spectrum Educational Supplies.
  - **Maxifun** – a large kit which allows for wheels and pulleys to be incorporated into designs. Can be purchased from First Class Ltd.

Figures 4.3, 4.4 and 4.5   ## ELECTRICAL COMPONENTS

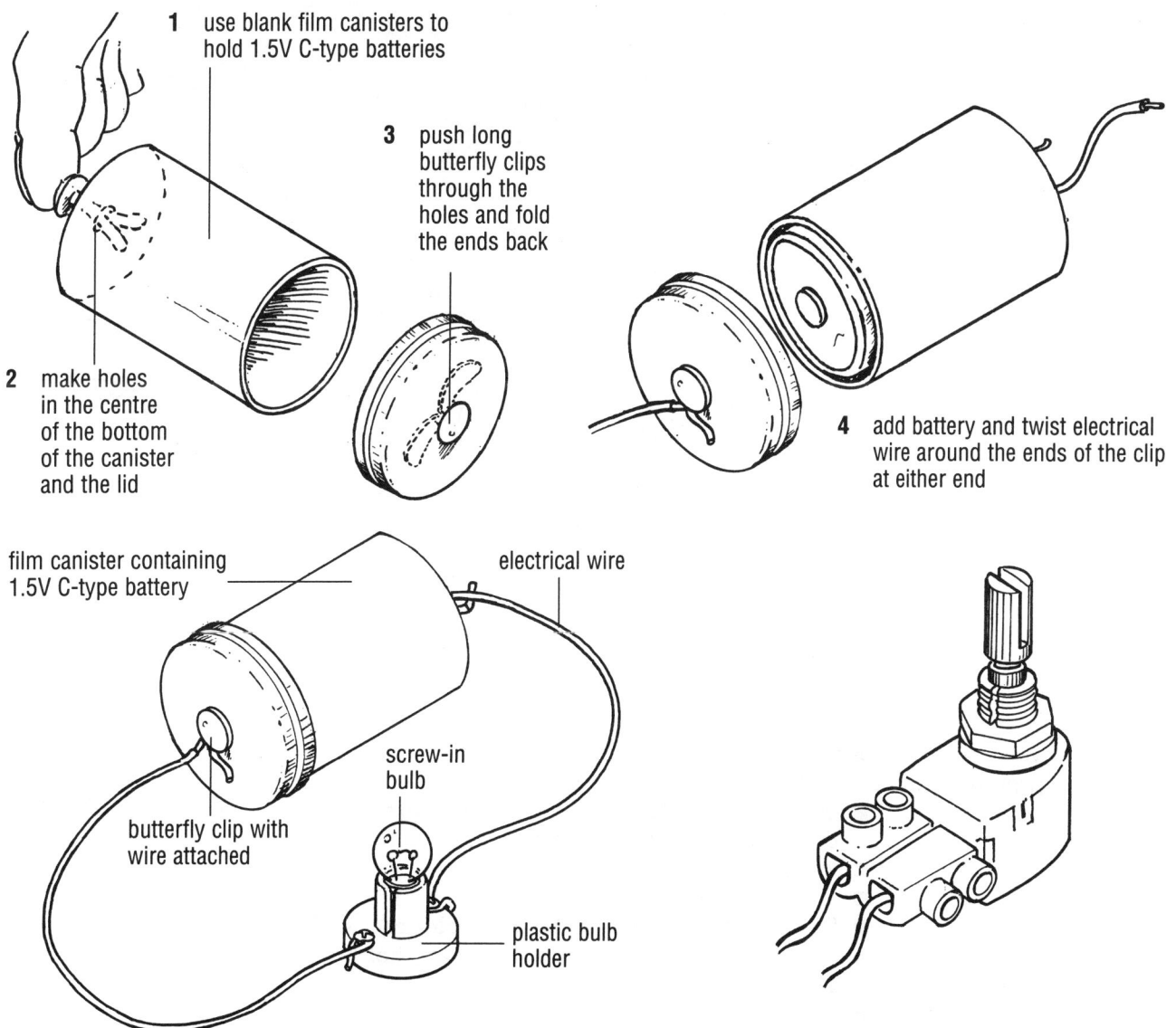

1 use blank film canisters to hold 1.5V C-type batteries

3 push long butterfly clips through the holes and fold the ends back

2 make holes in the centre of the bottom of the canister and the lid

4 add battery and twist electrical wire around the ends of the clip at either end

film canister containing 1.5V C-type battery

electrical wire

screw-in bulb

butterfly clip with wire attached

plastic bulb holder

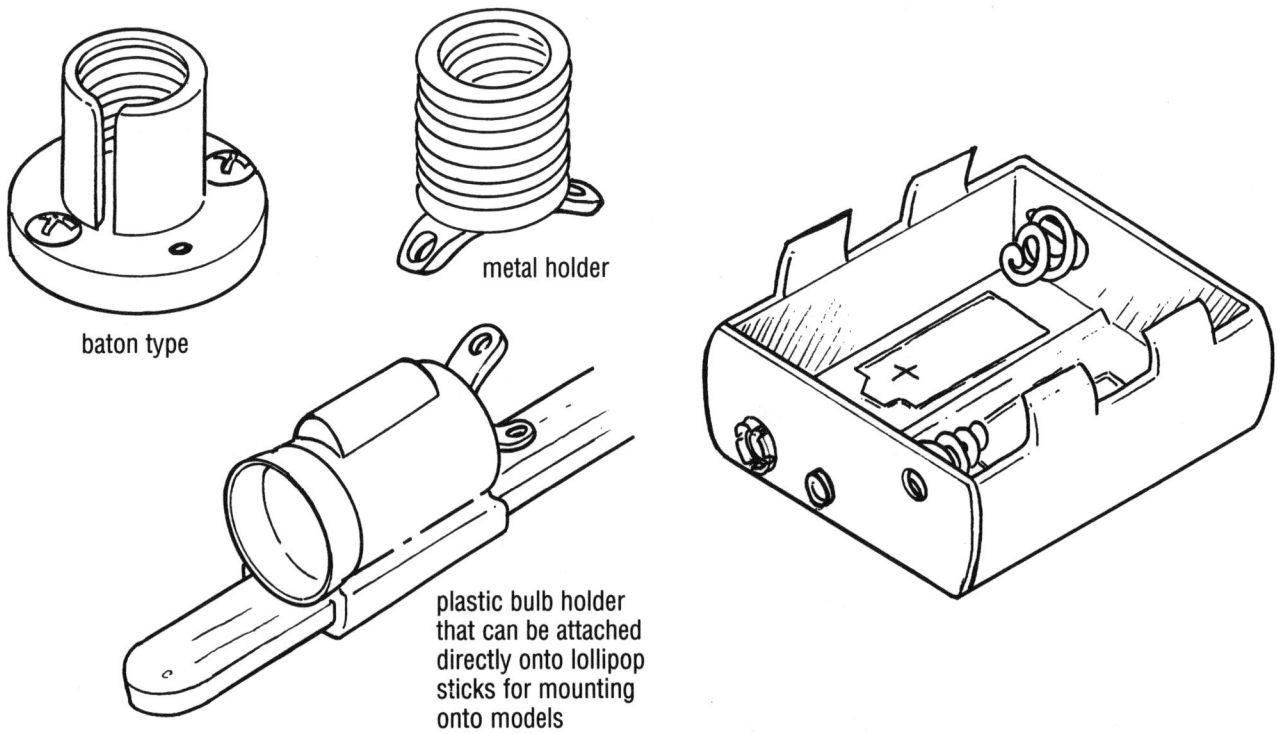

baton type

metal holder

plastic bulb holder that can be attached directly onto lollipop sticks for mounting onto models

Figures 4.6 and 4.7

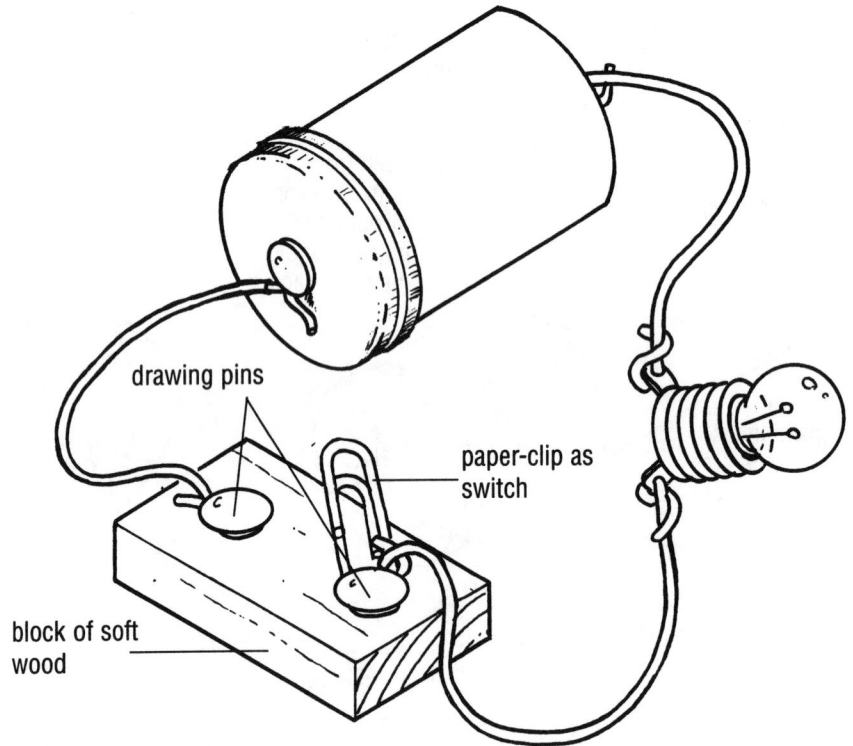

drawing pins

paper-clip as switch

block of soft wood

Figure 4.8

Equipment for circuitry work

- multi-core wire
- 2.5V and 3.5V round MES bulbs or 2.2V lens MES bulbs
- bulb holders – MES clip-on bulb holders are ideal and they slide directly on to lollipop sticks to aid mounting on to models.
- lollipop sticks (unused only)
- 1.5V 'C'-size batteries
- film canisters (for use as battery holders). See Figure 4.3
- commercial single and double battery holders – 'C' and 2'C' battery holders
- 3V to 6V buzzers
- 1.5V DC motors

- motor mounting clips
- propellers to fit on DC motor spindle
- paper fasteners
- aluminium foil
- speed control switch kit for dimming lights and varying the speed of a motor
- wire strippers
- plastic connector strips
- small flat-bladed screwdrivers
- crocodile clips – can be purchased complete with wire already attached

Equipment for introducing work on switches
- drawing pins
- paper clips
- small strips of soft wood for mounting switches
- commercial switches including press to make, press to break, press on/press off, slide and reed
- small magnets

Figure 4.9

simple switches

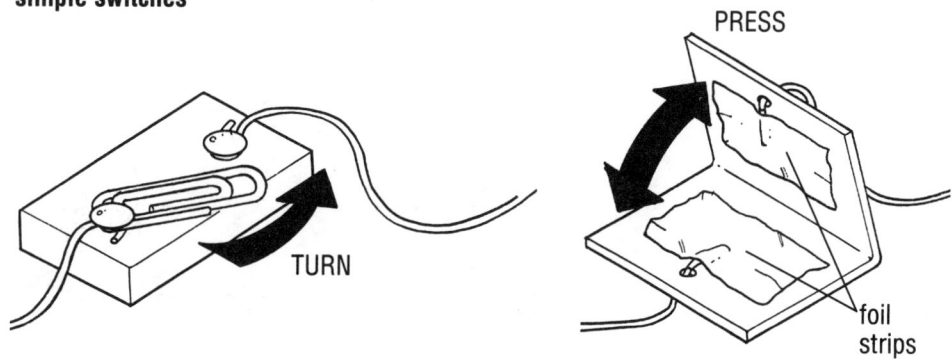

TURN

PRESS

foil strips

more complex switches

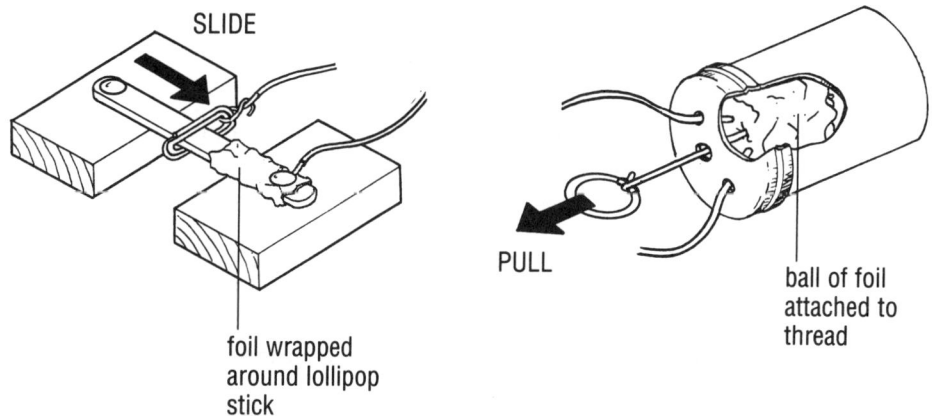

SLIDE

foil wrapped around lollipop stick

PULL

ball of foil attached to thread

Figures 4.10 and 4.11

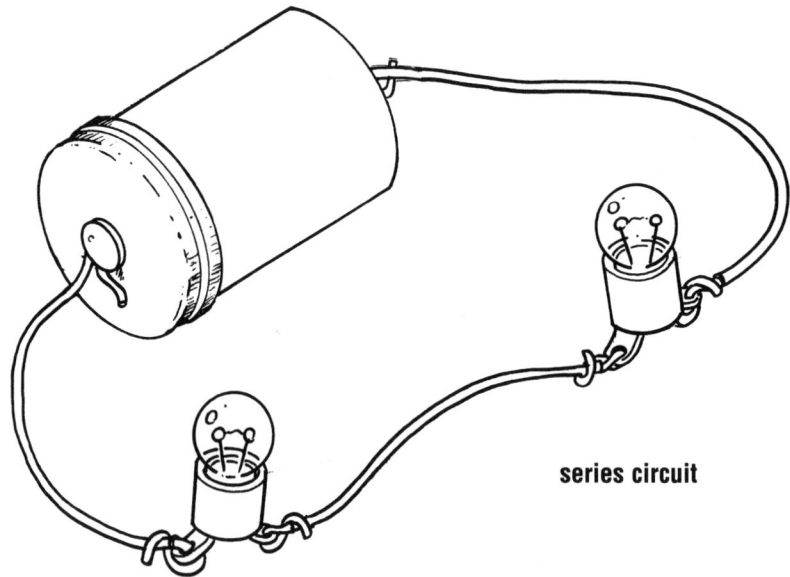

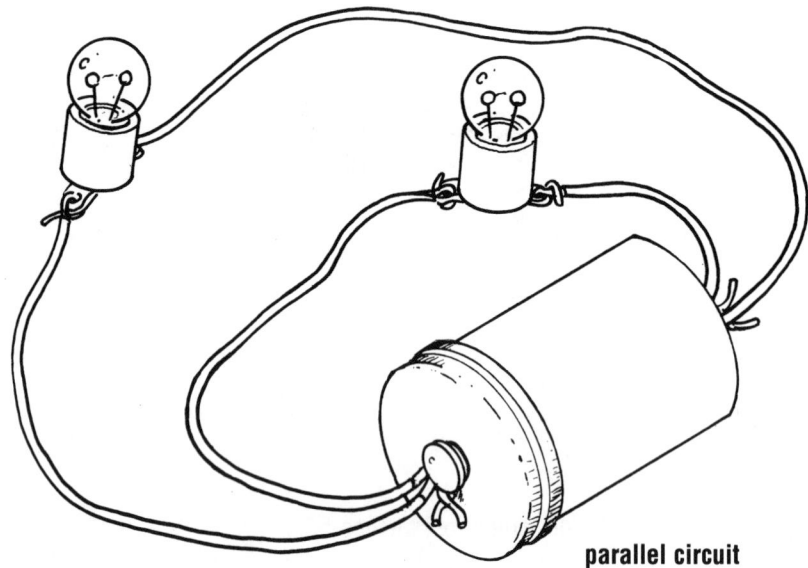

series circuit

parallel circuit

MECHANICAL COMPONENTS

A wide range of mechanisms can be created from card, including springs, spirals, coils and flaps.

■ Sliding movement can be created by using card guides for a strip to slide through.

■ Use paper fasteners (split pins) and a hole punch, to create a large enough hole for free movement, to create revolving and rocking movements.

■ Use paper fasteners, cotton lengths and tape to make pivots and levers to create flapping ears or arms for use on greeting cards or shadow puppets.

Pneumatics and hydraulics
■ 5 ml, 10 ml and 20 ml syringes
■ 3 mm clear plastic tubing
■ 2-way and 3-way valves
■ selection of small magnets – ring, small round and small square
■ tee connectors
■ straight connectors
■ balloons

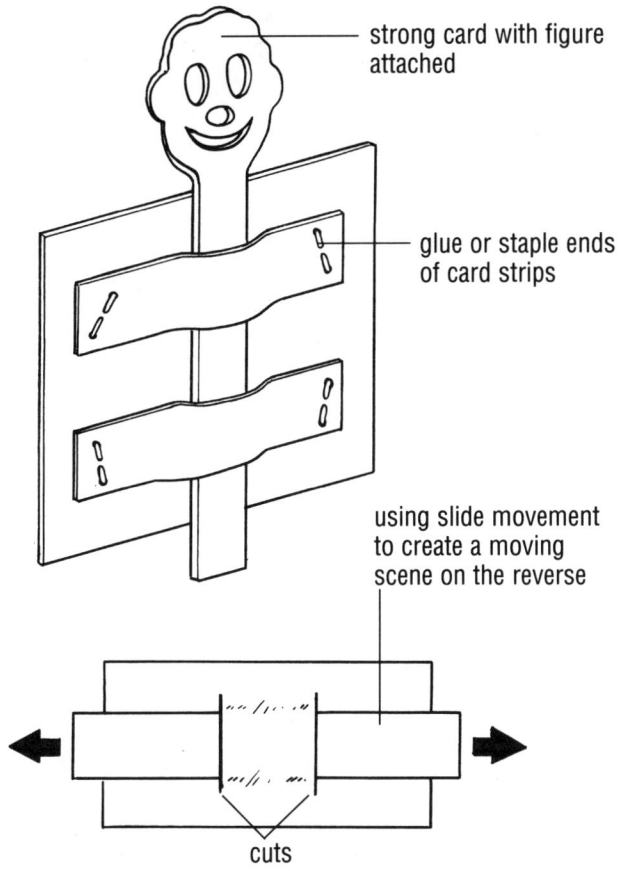

strong card with figure attached

glue or staple ends of card strips

using slide movement to create a moving scene on the reverse

cuts

Figure 4.12

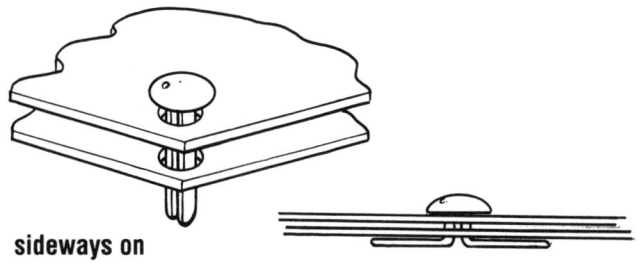

sideways on

from the front

Figure 4.13

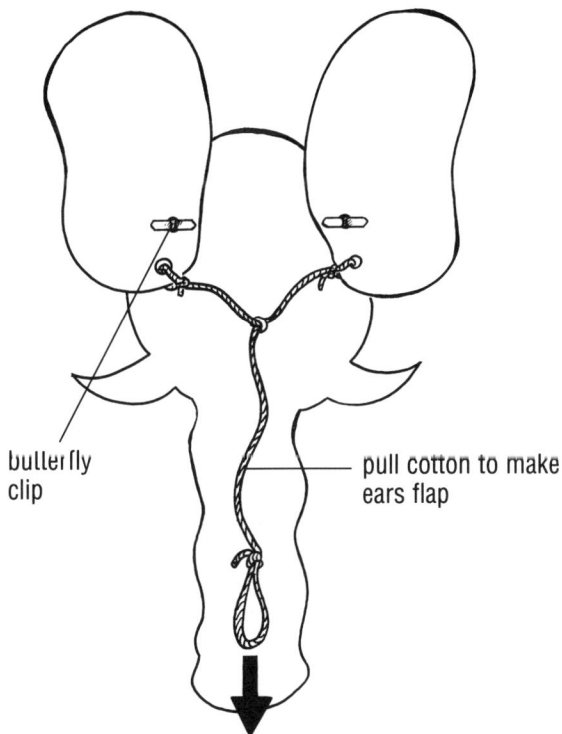

butterfly clip

pull cotton to make ears flap

Figure 4.14

Figure 4.15

Figure 4.16

balloon taped to tubing

5ml syringe

3-way
adaptor

plastic tubing

20ml syringe

Syringes of different sizes,
plastic tubing and balloons
can be combined to create
movement. The system can
be incorporated into models
to create, for example, the
opening of a crocodile jaw or
the eyes popping out of a
cardboard box robot.

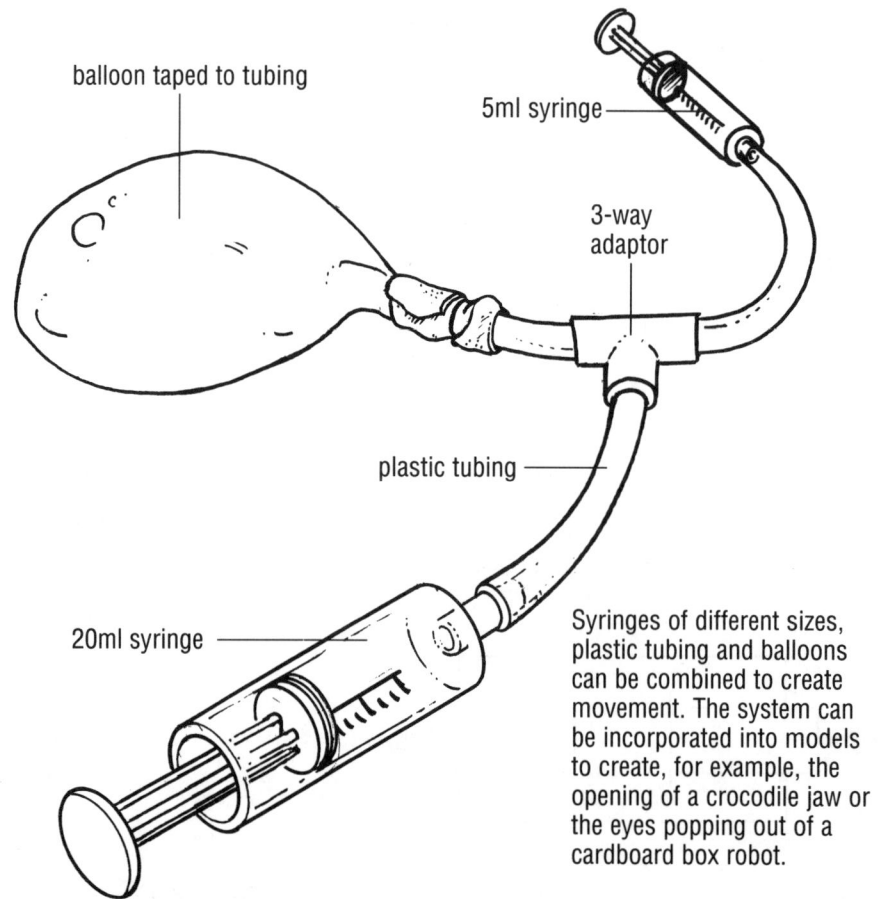

Gears and pulleys   Valiant gears and pulleys are an ideal set to purchase and are available in sizes that fit directly on to dowel or motors. Starter sets are available from suppliers.

Making gears
Making gears and incorporating them into children's models is not advised, as it takes a great deal of skill and time to construct a working mechanism.

When working with gears, provide a suitable kit like Start Gear or Lego Technics, whichever is best suited to what the children are making. When the children have made their mechanisms, look for ways in which they can be mounted on to their work, placed hidden away at the rear or 'boxed' in with a frame of card, wood, a box or parts from a construction kit.

Figure 4.17

Making pulleys
Pulleys can be made by sticking three wooden wheels together, the middle one being smaller. They can also be made by cutting two equal-sized circles out of card and sticking them on either side of a smaller-sized wooden wheel. Remember to punch a hole in the centre of the card before you stick it on to the side of the wooden wheel.

join three pre-cut wooden wheels together

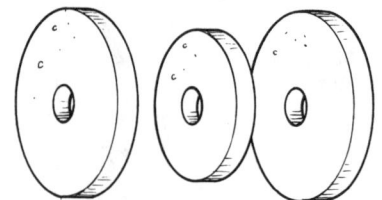

stick a circle of card to either side of a smaller wooden wheel

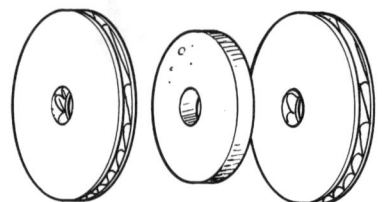

Figure 4.18

Cotton reels can be used as drum pulleys. Reduce the size of the diameter of the centre by inserting a length of 8 mm plastic tubing, which acts as a reducing sleeve. A length of 5 mm dowel can then be inserted and the cotton reel will then be fixed to the dowel.

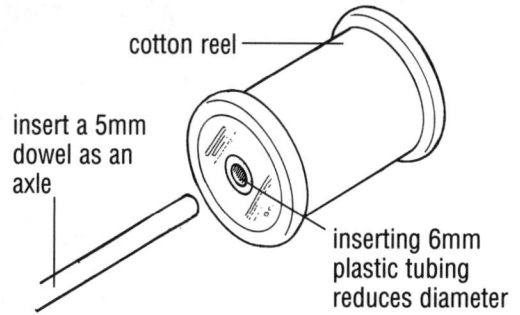

cotton reel

insert a 5mm dowel as an axle

inserting 6mm plastic tubing reduces diameter

Cams

Lego Technic kits are ideal for exploring cams and their uses. Activity cards can be purchased for each Lego Technic kit. Allow the children to construct the mechanism from the cam activity card and explore ways of varying the movement created by the cam. Their models may be incorporated into their work in the same way as outlined above under 'Making gears'.

Figure 4.19

cams can be used to create interesting forms of movement

by incorporating a cam into the design of this activity centre the face can be made to move up and down in the window

turn

face attached to dowel

dowel supports

dowel rests on top of cam

cam

wooden wheel for turning

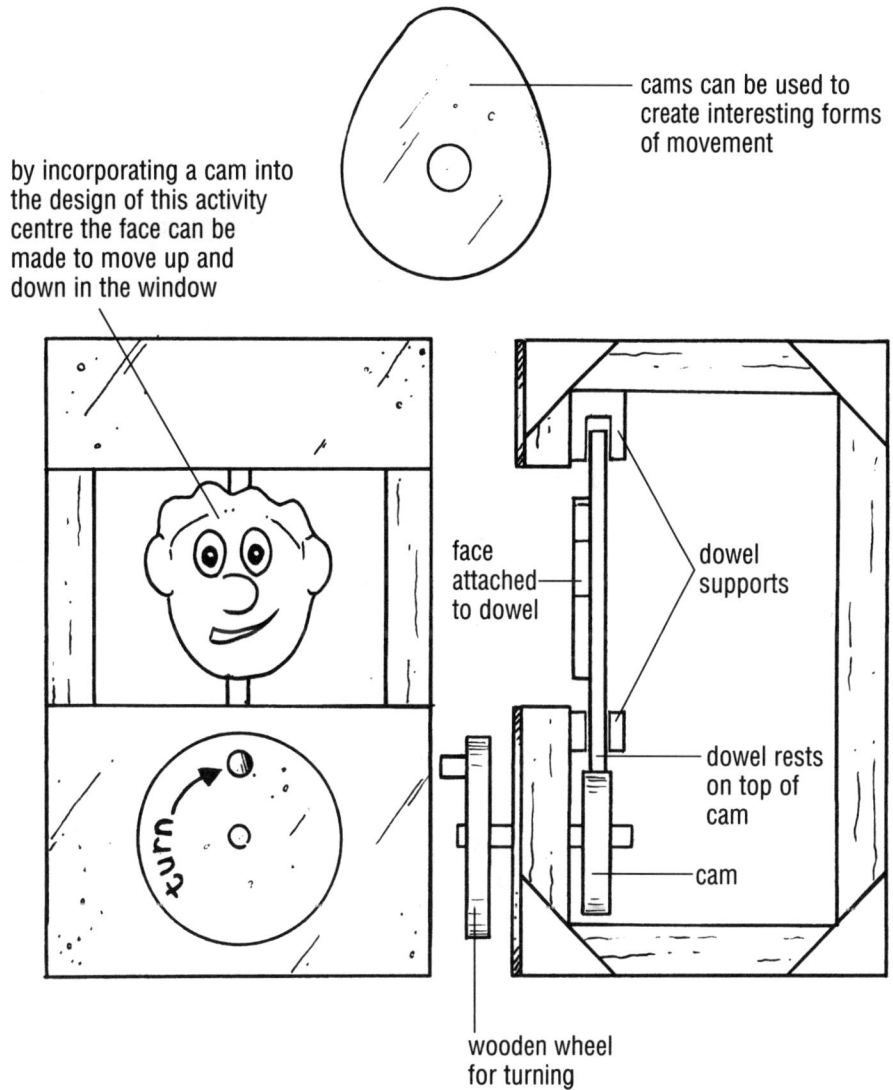

Wooden wheels with off-set centres can be made for use as cams, producing interesting forms of movement.

Information Technology

The following is not a definitive list of IT resources but includes equipment that supports work within the units of work set out in Section 3 'A Scheme of Work'.

COMMUNICATION AND DATA HANDLING

Word processor for:
- producing text to a given area, size and typeface for greeting cards, posters and packaging
- writing letters to companies or local businesses, seeking information about specific products.

Tape recorder for attaching to computer control systems to provide music or messages to accompany an environment or object that the children have made.

Graphics program for producing:
- patterns for wrapping paper or to cover packages
- drawings as part of a child's design, e.g. for a book
- a logo for a product, to be placed on letter paper or packages.

Desk top publishing package for producing:
- newspapers, pamphlets and leaflets for advertising a product
- a guide sheet for a place the children have visited
- any leaflet that requires a mixture of text and graphics.

Clip art files for providing children with a variety of pictures, labels and text on a variety of subjects for producing letters, posters or greeting cards.

Database for:
- gathering information on a variety of products so that comparisons can be made that inform children's designs
- collection of information gathered from a survey or questionnaire that the children have undertaken.

CONTROLLING

Computer control kit (control interface) for:
- writing simple procedures to work lights, buzzers and motors
- creating an interesting light sequence for inclusion in an illumination the children have made. They could create a 'Mexican wave' sequence where the lights seem to move backwards and forwards or a 'snake' sequence where the lights appear to move along in a line
- controlling an illumination so that it lights up when it is dark
- controlling any illuminated sign so that it lights up when it is dark and goes off when it is light.

MODELLING

CD Rom for researching how things work or have been made.

Bibliography

Design and Technology: A review of inspection findings 1993/4 by Office for Standards in Education (Ofsted), (HMSO)

Design and Technology – Primary Guidance by the Design and Technology Association (DATA)

Electricity and Magnetism: A Guide for Teachers: Knowledge and Understanding Series by School Curriculum and Assessment Authority (SCAA)

Implementing Design and Technology at key stages 1 and 2 by School Curriculum and Assessment Authority (SCAA)

Make It Safe: Safety guidance for the teaching of Design and Technology at Key Stages 1 and 2 by the National Association of Advisers and Inspectors in Design and Technology (NAAIDT)

Move it — with water and air (Technology Teaching Systems)

The Electricity Book 1 (Technology Teaching Systems)

The Electricity Book 2 (Technology Teaching Systems)

The Way Things Work by David Macaulay (Dorling Kindersley)

Working with Food in Primary Schools by Jenny Ridgewell (Ridgewell Press)

Useful addresses

Design and Technology
Association (DATA)
16 Wellesbourne House
Walton Road
Wellesbourne
Warwickshire CV35 9JB

First Class Ltd.
Imperial House
Willoughby Lane
London N17 0SP
Tel: 01885 3311

G.C. Products
New Bungalow
Northbeck
Scredington
Lincolnshire

Heron Educational
Carrwood House
Carrwood Road
Chesterfield
Derbyshire
Tel: 01246 453354

Spectrum Educational Supplies
Tudor Lodge
Babworth
Retford
Nottinghamshire
Tel: 01777 708306

Technology Teaching Systems
Unit 4
Park Road
Holmewood
Chesterfield S42 5UY
Tel: 01246 850085

Appendix 1 – Unit planning sheet

Year/s	Term	Time allocation
Unit of work		
Title		

Context

Outcome

Cross-curricular links

RESOURCES FOR INVESTIGATING

RESOURCES FOR MAKING

ACTIVITIES   Focused practical task

Investigative, disassembly and evaluative activities

PLANNING PRIMARY DESIGN AND TECHNOLOGY   © John Murray

Design and make assignment

EXTENSION ACTIVITIES

ASSESSMENT   The following types of responses may be observed as children work through this unit.

at Level __   **Designing**

Making

at Level __   **Designing**

Making

at Level __   **Designing**

Making

© John Murray   *PLANNING PRIMARY DESIGN AND TECHNOLOGY*

KEY STAGE 1 MAPPING SHEET

Category	Item						
Knowledge and understanding	g. vocabulary						
	f. health and safety						
	e. quality						
	c/d. products and applications						
	b. structures						
	5a. mechanisms						
Making skills	f. evaluate their products as they are developed, identifying strengths and weaknesses						
	e. make suggestions about how to proceed						
	d. apply simple finishing techniques						
	c. assemble, join and combine materials and components						
	b. measure, mark out, cut and shape						
	4a. select tools, materials and techniques						
Designing skills	f. consider their design ideas as these develop and identify strengths and weaknesses						
	e. make suggestions about how to proceed						
	d. develop and communicate design ideas through freehand drawing and modelling						
	c. develop their ideas through shaping, assembly and rearranging materials and components						
	b. clarify their ideas through discussion						
	3a. draw on their own experience to help generate ideas						
Materials	construction kits						
	food						
	textiles						
	reclaimed materials						
	2a. sheet materials						
	Unit of work						

PLANNING PRIMARY DESIGN AND TECHNOLOGY

KEY STAGE 2 MAPPING SHEET

Category	Unit of work						
Knowledge and understanding — k. vocabulary							
j. health and safety							
h./i. quality							
f./g. products and applications							
e. structures							
d. control – electrical							
c. control – mechanical							
b. materials and components – combining and mixing to make more useful							
5a. materials and components – working characteristics relate to their use							
Making skills — g. implement improvements they have identified							
f. evaluate and test their product							
e. select and plan use of materials, equipment and processes							
d. apply additional finishing techniques							
c. join and combine materials and components							
b. measure, mark out, cut and shape							
4a. select appropriate tools, materials and techniques							
Designing skills — g. evaluate design ideas against user and purpose and suggest ways forward							
f. develop a planned sequence and suggest alternatives							
e. communicate and model ideas							
d. consider appearance, function, reliability							
c. clarify ideas, develop criteria and suggest ways forward							
b. generate ideas, considering the users and purposes							
3a. use information sources in their designing							
Materials — construction kits							
mechanical components							
electrical components							
food							
textiles							
mouldable materials							
framework materials							
2a. stiff and flexible sheet materials							

Appendix 4

Tasting Food

Sample	☹	🙁	😐	🙂	😊

Sample	☹	🙁	😐	🙂	😊

PLANNING PRIMARY DESIGN AND TECHNOLOGY

© John Murray

Appendix 5

Star Diagram

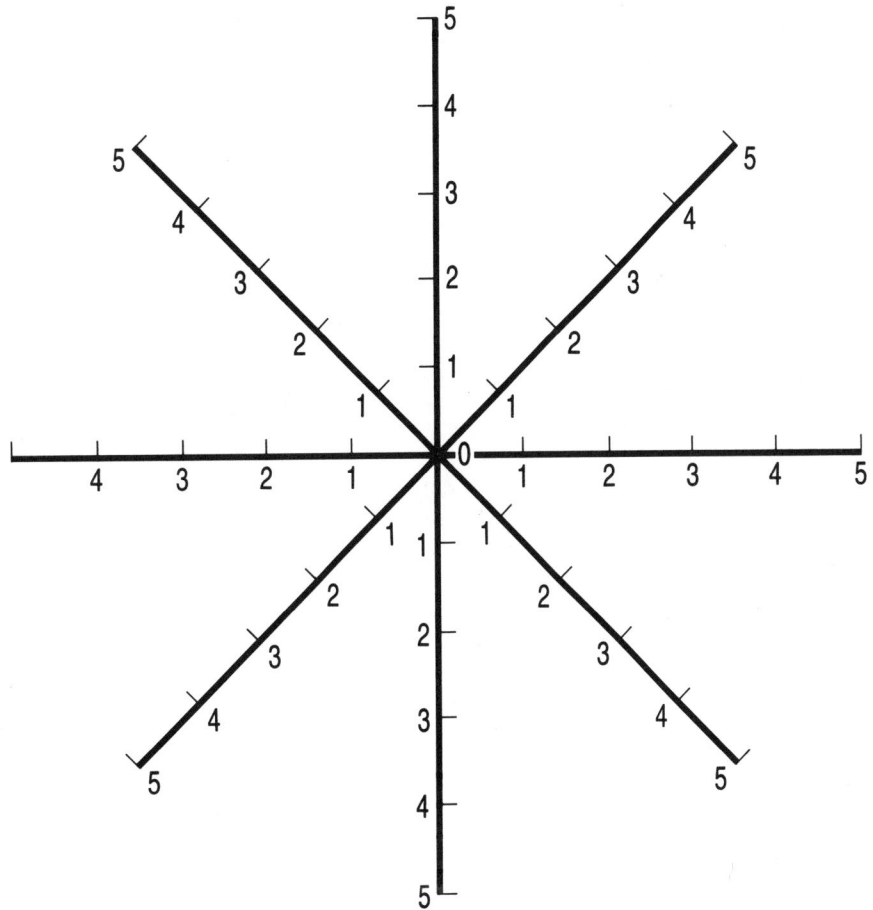

0 = not at all

3 = it's OK

5 = very

What are you testing? _____

What have you learnt? _____

PLANNING PRIMARY DESIGN AND TECHNOLOGY